T0265417

THE (1+1)-NONLINEAR UNIVERSE OF THE PARABOLIC MAP AND COMBINATORICS

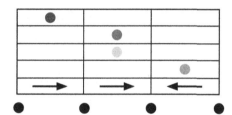

THE (1+1)-NONLINEAR UNIVERSE OF THE PARABOLIC MAP AND COMBINATORICS

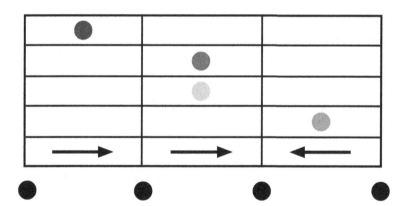

James D. Louck · Myron L. Stein

World Scientific

NEW JERSEY · LONDON · SINGAPORE · BEIJING · SHANGHAI · HONG KONG · TAIPEI · CHENNAI

Published by

World Scientific Publishing Co. Pte. Ltd.

5 Toh Tuck Link, Singapore 596224

USA office: 27 Warren Street, Suite 401-402, Hackensack, NJ 07601

UK office: 57 Shelton Street, Covent Garden, London WC2H 9HE

Library of Congress Cataloging-in-Publication Data
Louck, James D.
   The (1+1)-nonlinear universe of the parabolic map and combinatorics / by James D. Louck (Retired Los Alamos National Laboratory Fellow, USA), Myron L. Stein.
      pages cm
   Includes bibliographical references and index.
   ISBN 978-9814632416 (hardcover : alk. paper)
   1. Nonlinear theories. 2. Chaotic behavior in systems. 3. Combinatorial analysis. 4. Mathematical analysis.
I. Stein, M. L. II. Title. III. Title: Nonlinear universe of the parabolic map and combinatorics.
   QA427.L68 2015
   511'.6--dc23

                                              2014042607

British Library Cataloguing-in-Publication Data
A catalogue record for this book is available from the British Library.

Printed in Singapore

*To the memory of Robert L. Bivins, Nicholas C. Metropolis,
and Myron L. Stein, without whom this monograph
could not have been completed.*

Preface

The original motivation for this monograph was to set forth the early contributions from the Theoretical Division at Los Alamos National Laboratory to the foundations of chaos theory. Overviews of work done up to 1983 have already been given in LA–2305,1959 and in LA–9705,1983, which are available electronically on request from the Laboratory. These reports remark on the foundations of the subject as set forth in early papers by Stein and Ulam [1], N. Metropolis *et al* [2-3], Feigenbaum [4-9, 12], Feigenbaum *et al* [10], Beyer and Stein [11], Beyer *et al* [13], Stein [14], and the book by Bivens *et al* [15]. These are the primary references leading to the viewpoints developed in this monograph. The evolution of ideas beginning with the above references is an important ingredient of this monograph. It is this aspect that is focused on in the Preface, but this is intertwined by a preview of a major shift in viewpoint that developed as the writing progressed.

Principal properties promoted and developed by Bivins *et al* [16] are those of the inverse graph, which for a general function f with real values $f(x)$ is a collection of single-valued complex functions called branches. For the case at hand, the basic function is the parabola p_ζ, which is defined by its set of values $p_\zeta(x) = \zeta\, x(2-x)$, $x \in (-\infty, \infty)$. The parameter ζ is, for the most part, taken to be real with values of ζ in the closed interval $\zeta \in [0,2]$. (It turns out, however, that all real values $\zeta \in (-\infty, \infty)$ are important.) This method based on properties of the inverse graph was itself motivated by the discovery that the inverse graph had the property of being sometimes complex and sometimes real, but with the extraordinary property that each such inverse function becomes real at a characteristic value of $\zeta \in [0,2]$, and remains real for all greater values of ζ. Thus, a theory emerged that was based on function composition, one that also allowed the creation of objects such as curves and fixed points.

The major shift in viewpoint occurred when an algorithm was discovered during the write-up of the monograph that allowed the generation of the inverse graph for $n-1$ to n. This placed the subject clearly in the arena of a complex adaptive system, where a complex adaptive system is taken to be a system whereby a few principal axioms lead to a system rich in structure and predictive power. For the problem at hand, this was realized by some simple implementable rules, ones that could also be calculated numerically and verified visually. Thus, the idea of an algorithmic-computer-generated inverse graph had evolved that fits well with the notion of a complex adaptive system. But what about applications and predictability?

The complex adaptive system viewpoint is further enriched by properties of the inverse graph that can be interpreted in terms of combinatorial concepts such as a total order relation on all branches of the inverse graph that exist at a given value of ζ, an order relation that is never violated, up to and including all positive values of ζ. Moreover, this labeling of branches of the inverse group can be realized by hook tableaux, which are special Young standard tableaux, or, equivalently, by special Gelfand-Tsetlin patterns. Such patterns can be realized as isotropic quantum oscillators.

The complex system applications do not end here; they continue still into biology and beyond: see Bell *et al* [20] and Bell and Torney [21] with yet further applications to Galois groups by Byers and Louck [23] and to Conway numbers by Byers and Louck [24-25].

Most importantly for this monograph the issue of an application to General Relativity arises based on the mathematical operation of function composition; the case for a complex adaptive system has been established. Whether or not it provides any meaningful insights into General Relativity remains to be seen. The authors have no experience working in General Relativity other than a general introduction, which is inadequate for such judgments. But there is still an obligation to point out the possibilities. It appears that the existence of **A Fully Deterministic Chaos Theory** is a **basic property** with a potential application to General Relativity.

The first author takes full responsibility for the viewpoints presented in this monograph. It is, of course, the case that these viewpoints could not have emerged without the extraordinary interaction between computer calculations and the development of theory.

A somewhat unusual style of presentation has been utilized in this monograph. Many pictures of inverse graphs at various parameter-values $\zeta_1 < \zeta_2 < \cdots < \zeta_t < \cdots$ are given that illustrate crucial properties of the ζ-parameter evolution of the inverse graph. Thus, the notion that the system under study is a complex adaptive system is re-enforced by computer calculations in which the inverse graph exhibits the predicted properties. Sufficiently many computer graphs are included, as needed to exhibit a particular property. For a vivid mental picture, it is often useful to think of ζ as time. It is in this time-evolution of the n-th iterate of the inverse graph that the classification by words on two letters comes into play, their fundamental role being to enumerate the branches of the inverse graph. The patterns exhibited by explicit computer computations of the shape of graphs and the expression of their explicit mathematical forms is a nice example of how one mode of presentation generates and re-enforces insights into the other. This accounts for the dedication of this work to the memories of R. L. Bivins, Nicholas C. Metropolis, and Myron L. Stein. World Scientific graciously allowed the inclusion of Myron's name on the cover, since his computational contribution was completed before his death. It is quite impossible to express the compassion and support of Editor Lai Fun Kwong.

The organization of this work, the many pictures of the inverse graph aside, is quite standard, as detailed in the **Contents**. It is emphasized that this monograph is far too technical and detailed to be a textbook. It is intended for readers with a perchance for the unusual and unexpected. Most will probably have a background in physics or mathematics.

This work could not have been completed without 54 years of enduring patience and endearing love of my wife Marge and the expert computer maintenance support of our son Tom. Thanks are given to David C. Torney, Peter W. Milonni, and Michael M. Nieto (deceased 2013) for many useful discussions on the foundations of mathematics and physics. Also, thanks to Librarians Michelle Mittrach and Kathy Varjabedian who diligently provided electronic copies of references. The viewpoints and attributions expressed herein are mine alone.

James D. Louck

Contents

Chapter 1

Introduction and Point of View

In this opening chapter, a synthesis is given of the results found in Refs. [1-5, 15-19]. The ideas, procedures, and definitions introduced in this Chapter are drawn from these references. Slight variations in notations may occur. The idea of this overview is to capture many of the over-riding features of the so-called ζ-evolution of the various graphs without giving all the many details needed for their complete description.

1.1   Function Composition and Graphs

The principal mathematical operation that produces most curves generated and discussed in this monograph is the operation on pairs of functions known as *composition*. The composition of a pair of functions f and g is denoted by $f \circ g$. It is defined by giving its value, denoted $(f \circ g)(x)$, in terms of the values of the functions f and g, as expressed by

$$(f \circ g)(x) = f\Big(g(x)\Big). \tag{1.1}$$

Thus, $(f \circ g)(x)$ is the value of $f(x)$ at $x = g(x)$. The operation of composition is noncommutative, but associative:

$$f \circ g \neq g \circ f; \quad (f \circ g) \circ h = f \circ (g \circ h), \tag{1.2}$$

as verified directly from the definition (1.1).

The composition of pairs of functions generalizes directly to that of the composition of arbitrarily many functions:

$$(f_1 \circ f_2 \circ f_3)(x) = f_1\bigg(f_2\Big(f_3(x)\Big)\bigg),$$

$$(f_1 \circ f_2 \circ f_3 \circ f_4)(x) = f_1\bigg(f_2\Big(f_3\big(f_4(x)\big)\Big)\bigg), \tag{1.3}$$

$$\vdots$$

$$(f_1 \circ \cdots \circ f_{n-2} \circ f_{n-1} \circ f_n)(x) = f_1\left(\cdots f_{n-2}\left(f_{n-1}\left(f_n(x)\right)\right)\cdots\right).$$

Because the rule of composition is associative, no additional parenthesis pairs are needed in the left-hand side of these relations. There are n parenthesis pairs ( ) on the right-hand side: n left parentheses (one following each f_i, and each matched with a right parenthesis), thus constituting a parenthesis pair ( ), where all n right parentheses occur in succession at the right-most end of each of relations (1.3).

The inverse of a function f with values $f(x)$ is denoted by f^{-1} and is defined here to be a single-valued function with values denoted by $f^{-1}(x)$ such that

$$f\left(f^{-1}(x)\right) = f^{-1}\left(f((x)\right) = x. \tag{1.4}$$

Thus, the inverses to f are solutions of the equation $f(y(x)) = x$, and in general there can be several distinct solutions; careful attention must be paid to the domains of definition of f and f^{-1}. In this monograph, distinct inverses to a given single real-valued function f are called **branches.** An inverse f^{-1} to f can also be defined by the composition rule $f^{-1} \circ f = f \circ f^{-1} = I$, where I is the identity function with values $I(x) = x$. The interest here is not with all the subtleties that arise in considering collections of functions and their compositions, but, rather, with the properties of the n-fold composition of a single function — the parabola defined by

$$p_\zeta(x) = \zeta x(2 - x), \ \zeta \in (0, \infty); \ x \in (-\infty, \infty). \tag{1.5}$$

Most of the interest of the present monograph is directed toward the development of the properties of the 2^n-fold compositions of the two branches of the inverse function to $p_\zeta(x)$ as defined by

$$\Phi_\zeta(1; x) \ = \ 1 + \sqrt{1 - \frac{x}{\zeta}}; \ \ \Phi_\zeta(-1; x) = 1 - \sqrt{1 - \frac{x}{\zeta}}, \tag{1.6}$$

$$\zeta \in (0, \infty), \ x \in (-\infty, \zeta).$$

Each of these branches is, of course, a real single-valued function of x in the domain $x \in (-\infty, \zeta)$, and the two functions join smoothly at $x = \zeta$ to constitute what will be called a p-curve. A p-curve is the joining of two branches as illustrated in the following schematic picture for the (x, y)-planar graph of the branches $\Phi_\zeta(1; x)$ and $\Phi_\zeta(-1; x)$ for $x \in (0, \zeta]$:

This picture depicts a **right-moving** p-curve with increasing ζ. The general polynomials of interest are the real polynomials of degree 2^n in x defined by the n-fold composition of p_ζ:

$$p_\zeta^n(x) = \Big(p_\zeta \circ p_\zeta \circ \cdots \circ p_\zeta\Big)(x) = p_\zeta\Big(\cdots \Big(p_\zeta\big(p_\zeta(x)\big)\Big)\cdots\Big), \qquad (1.8)$$

where there are n parenthesis pairs in this expression for an n-fold composition of one and the same parabola function p_ζ. It is very important to observe that the parameter ζ is fixed at the same value in the composition (1.8). Thus, while it is allowed that ζ be any value $\zeta \in (0,\infty)$, the operation of composition is to be effected only for specified ζ in its domain of definition, as illustrated by

$$\begin{aligned} p_\zeta^2(x) &= \Big(p_\zeta \circ p_\zeta\Big)(x) = \zeta\, x(2-x)\Big|_{x=\zeta\, x(2-x)} \\ &= \zeta^2\, x(2-x)\Big(2 - \zeta\, x(2-x)\Big). \end{aligned} \qquad (1.9)$$

A very useful rule satisfied by such compositions is:

$$p_\zeta^n(x) = \Big(p_\zeta^{n-m} \circ p_\zeta^m\Big)(x) = p_\zeta^{n-m}\Big(p_\zeta^m(x)\Big),$$

$$m = 1, 2, \ldots, n-1, \qquad (1.10)$$

$$p_\zeta^1(x) = p_\zeta(x) = \zeta\, x(2-x).$$

The n-fold iterate $p_\zeta^n(x)$ of $p_\zeta^1(x)$ is a polynomial of degree 2^n in the variable x and degree $2^n - 1$ in the parameter ζ. Thus, the polynomial is of the form

$$p_\zeta^n(x) = \sum_{k=0}^{2^n} a_k^{(n)}(\zeta)\, x^{2^n - k}, \qquad (1.11)$$

where the coefficients are real polynomials in the parameter ζ with leading coefficient $a_0^{(n)}(\zeta) = 2^n - 1$ and successive coefficients of lower degree. A recurrence relation for the polynomials is given by

$$p_\zeta^n(x) = (p_\zeta^{n-1} \circ p_\zeta^1)(x) = p_\zeta^{n-1}\Big(p_\zeta^1(x)\Big). \qquad (1.12)$$

Thus, an explicit recurrence for the coefficients $a_k^{(n)}(\zeta)$ themselves can be obtained, if desired, by combining relation (1.12) and (1.11) with the appropriate relations from (1.10). The main point is: *The polynomials $p_\zeta^n(x)$ are uniquely defined for all positive n.*

The graph H_ζ^n of interest is defined as the set of points in the Cartesian plane $\mathbb{R}^2$ given by

$$H_\zeta^n = \Big\{\big(x,\, p_\zeta^n(x)\big)\Big| x \in [0,\,\infty)\Big\}, \quad \zeta \in (0,\infty). \qquad (1.13)$$

Many of the interesting features of this graph make their appearance for
$x \in [0, 2]$, although other domains, even including negative x, are of interest.
A principal feature of all graphs presented in Chapters 5-7 is that they are
presented at a value of the parameter ζ that is **specified** (fixed). The values
of x then determine the basic **shape** of the underlying curve in the (x, y)-
plane for the specified value of ζ; this set of real points constitute the graph
H^n_ζ: It is a continuous smooth curve (all derivatives exist at all points) in
$\mathbb{R}^2$. This set of points is also called the *shape of the graph H^n_ζ at ζ.*

As the parameter ζ changes continuously, the shape of the curve, which
is denoted by $\mathcal{H}^n_\zeta$, changes smoothly. In particular, the change in shape for
increasing ζ is called the ζ-*evolution* of the curve (or graph). Indeed, it is
often very useful to think of ζ as a time-like parameter; hence, the shape
$\mathcal{H}^n_\zeta$ is a "snapshot" of the graph at a given time, and the ζ-evolution is the
nonlinear time progression of the graph. The ζ-evolution of the graph is
unexpectedly elegant, expressing its unfolding shape in terms of the creation
of new "subcurves" and their symmetry. It is the purpose of this monograph
to give its description for all n.

There is a simple underlying reason for the origin of the features appearing
in the ζ-evolution of the graph H^n_ζ: *This is revealed in the structure of the
inverse graph.* If $H_f = \{(x, f(x)) \mid x \in D_f\}$ is the graph of a real single-
valued function f with values $f(x)$ and domain of definition $x \in D_f \subseteq$
$\mathbb{R}$, then, by definition, the set of points $H_{f^{-1}} = \{(x, f^{-1}(x)) \mid x \in D_{f^{-1}}\}$,
where $D_{f^{-1}} \subseteq \mathbb{R}$ is the domain of definition of the branch f^{-1}, constitutes a
subgraph of the inverse graph. But there is such an inverse graph for each
distinct inverse function f^{-1} of f; hence, it is the union $\cup_{f^{-1}} H_{f^{-1}}$ over all
distinct inverse subgraphs that constitutes the full inverse graph to H_f. This
simple description of the inverse graph holds unambiguously for the inverse
graph of the n-fold composition of the parabolic map $p_\zeta(x) = \zeta x(2 - x)$,
although care must be taken in defining the inverse function. In terms of
these notations, the graph H^n_ζ is given by

$$H^n_\zeta = H_{p^n_\zeta} = \left\{ (x, p^n_\zeta(x)) \,\Big|\, x \in [0, \infty) \right\}, \tag{1.14}$$

where the n-fold composition of the basic parabola $p^1_\zeta(x) = p_\zeta(x) = \zeta x(2-x)$
is defined in (1.8). It is the inverse graph to $H_{p^n_\zeta}$ that is sought for each
specified $\zeta \in (0, \infty)$. In terms of the present notations, the inverse graph
is denoted by $H_{f^{-1}_\zeta}$, where $f = p^n_\zeta$. For the case at hand, this somewhat
awkward notation is replaced by

$$G^n_\zeta = H_{f^{-1}_\zeta}\Big|_{f = p^n_\zeta}. \tag{1.15}$$

Thus, G^n_ζ denotes the inverse graph to the graph H^n_ζ. By definition:

The graph H^n_ζ and its inverse G^n_ζ are subsets of points in the real plane $\mathbb{R}^2$.

It is useful to illustrate the above definition of the inverse graph G^n_ζ before
proceeding to the general case.

Examples. The inverse graph G_ζ^n, $n = 1, 2$:

$n = 1$. The inverse graph is the union of two single-valued real branches:

$$G_\zeta^1(1) = \left\{ \left(x, \Phi_\zeta(1; x) \right) \,\middle|\, x \in [0, \zeta] \right\}, \ \zeta \in (0, \infty),$$

$$G_\zeta^1(-1) = \left\{ \left(x, \Phi_\zeta(-1; x) \right) \,\middle|\, x \in [0, \zeta] \right\}, \ x \in (\zeta, \infty); \quad (1.16)$$

$$G_\zeta^1 = G_\zeta^1(1) \cup G_\zeta^1(-1), \ \zeta \in (0, \infty).$$

These two branches join smoothly at their common point at $x = \zeta$. These two branches constitute a right-moving p-curve as ζ increases, as shown in (1.7).

$n = 2$. The inverse graph is the union of two single-valued real branches or of four single-valued real branches, depending on the value of ζ. The branches are given initially by the following definitions in terms of the four ($2^2 = 4$) ways of ways of composing the two square-root forms (1.6), even when some square-roots are complex:

$$\Phi_\zeta\left((1, -1); x\right) = 1 + \sqrt{1 - \frac{1}{\zeta} \Phi_\zeta(-1; x)}$$

$$= 1 + \sqrt{1 - \frac{1}{\zeta}\left(1 - \sqrt{1 - \frac{x}{\zeta}}\right)},$$

$$\Phi_\zeta\left((1, 1); x\right) = 1 + \sqrt{1 - \frac{1}{\zeta} \Phi_\zeta(1; x)}$$

$$= 1 + \sqrt{1 - \frac{1}{\zeta}\left(1 + \sqrt{1 - \frac{x}{\zeta}}\right)};$$

$$(1.17)$$

$$\Phi_\zeta\left((-1, 1); x\right) = 1 - \sqrt{1 - \frac{1}{\zeta} \Phi_\zeta(1; x)}$$

$$= 1 - \sqrt{1 - \frac{1}{\zeta}\left(1 + \sqrt{1 - \frac{x}{\zeta}}\right)},$$

$$\Phi_\zeta\left((-1, -1); x\right) = 1 - \sqrt{1 - \frac{1}{\zeta} \Phi_\zeta(-1; x)}$$

$$= 1 - \sqrt{1 - \frac{1}{\zeta}\left(1 - \sqrt{1 - \frac{x}{\zeta}}\right)}.$$

The sequences (σ_1, σ_2), each $\sigma_i = 1$ or -1, keep account of the $\pm$ signs in front of the square roots in these relations (see (1.6)). In order that the square-roots are a single number, even when complex, the convention $\sqrt{z} = \sqrt{r} e^{i\phi/2}$ for $z = r e^{i\phi}, r \geq 0, 0 \geq \phi < 2\pi$, is adopted, where, as always, the square-root $\sqrt{r}$ of a positive number r is always a positive number. This choice for $\sqrt{z}$ for z complex has no effect on the inverse graph construction of G_ζ^n, since *the rule for constructing this inverse graph is always that those quantities that appear under a square-root symbol $\sqrt{\ }$ are to be nonnegative real numbers.*

The four composition relations (1.17) can be written:

$$\Phi_\zeta(\sigma;x) \;=\; \Big(\Phi_\zeta(\sigma_1) \circ \Phi_\zeta(\sigma_2)\Big)(x) = \Phi_\zeta\Big(\sigma_1;\Phi_\zeta(\sigma_2;x)\Big), \qquad (1.18)$$

$$\sigma \;=\; (\sigma_1,\sigma_2), \text{ each } \sigma_i = 1 \text{ or } -1.$$

The generalization of relations (1.18) to arbitrary n is given by the following composition rule, which is an unambiguous rule for constructing all inverse graphs, including their unique σ_i labels:

$$\Phi_\zeta\Big((\sigma_1,\sigma_2,\ldots,\sigma_n);x\Big) = \Phi_\zeta\Big((\sigma_1,\sigma_2,\ldots,\sigma_{n-1});\Phi_\zeta(\sigma_n;x)\Big); \qquad (1.19)$$

$$\Phi_\zeta(\sigma_n;x) = 1 + \sigma_n\sqrt{1-\frac{x}{\zeta}}, \text{ each } \sigma_n = 1 \text{ or } -1.$$

A full description of the functions that enter into the inverse graph at each value of ζ is now completed by giving the domain of definition of each function defined by the composition rule (1.19). This domain of definition may be expressed by

$$\Phi_\zeta^{(l)}(\sigma;x) \le \Phi_\zeta(\sigma;x) \le \Phi_\zeta^{(r)}(\sigma;x), \;\; \sigma = (\sigma_1,\sigma_2, \ldots, \sigma_n), \qquad (1.20)$$

where $\Phi_\zeta^{(l)}(\sigma;x)$ and $\Phi_\zeta^{(r)}(\sigma;x)$ denote, respectively, the left and right extremal points of the composition function (1.19), where all functions appearing in (1.17)-(1.20) are **required to be real.**

The real relations (1.17)-(1.20) give the full description at each value of ζ of the inverse graphs that appear in the present approach to chaos theory via inverse graphs. But the most remarkable property is yet to come: There is a unique subclass of inverse graphs (1.17)-(1.20) for which the reality conditions hold; that is, in general, relations (1.17)-(1.20) admit both real and complex solutions, but imposing the conditions that each solution be real gives exactly **one solution.** It is this solution that is sought; it can clearly be called a **completely deterministic chaotic solution.** It is this property of being deterministic within the general framework of chaos theory that appears to single-out this solution as appropriate for many applications.

The proof that there exists a completely deterministic set of solutions as a subclass of inverse graphs is yet to be demonstrated — little has been revealed on how to recognize one: An explicit method of constructing each and every inverse graph at each value of ζ is needed. It turns out that this is a nontrivial task. Much of the remainder of this chapter develops the ideas necessary for the proof that such a subclass of inverse graphs exists, and that this set can be generated recursively and reproducibly by computer calculations. This then leads to the concept of a complex adaptive system.

For the purposes of this monograph, a **complex adaptive system** is defined to be:

A collection of objects in which the objects are the undefined elements of the theory for which there are definite rules (axioms) for combining the objects that leads to rich and unexpected properties of the objects themselves.

Here, in this monograph, this will be demonstrated only for the mathematical operation of function composition. It is a principal goal of this monograph to show that the collection of inverse graphs constitutes a complex adaptive system in the sense defined above. Towards this goal, it is necessary to identify properties of the inverse graph that lead to this important conclusion. This is set forth in the remainder of this Chapter 1, together with other supporting properties.

It is useful to illustrate how the above methodology works for $n = 2$:

$$\Phi_\zeta\big((\sigma_1, \sigma_2); x\big) = \Phi_\zeta\big(\sigma_1; \Phi_\zeta(\sigma_2; x)\big);$$

$$\Phi_\zeta^{(l)}\big((\sigma_1, \sigma_2); x\big) \geq \Phi_\zeta\big((\sigma_1, \sigma_2); x\big) \geq \Phi_\zeta^{(r)}\big((\sigma_1, \sigma_2); x\big); \qquad (1.21)$$

$$\sigma_1 = 1, -1; \ \sigma_2 = 1, -1, \text{ in each of the above relations.}$$

By definition, the domain of definition functions in this relation are real. This, in turn, implies that

$$x \leq \zeta \text{ and } \Phi_\zeta(\sigma_2; x) \leq \zeta, \text{ each } \sigma_2 = 1, -1. \qquad (1.22)$$

Since $p_\zeta\big(\Phi_\zeta(\sigma_2; x)\big) = x$, the condition $\Phi_\zeta(\sigma_2; 1) = \zeta$ requires that $p_\zeta(\zeta) = \zeta^2(2 - \zeta) = 1$; that is,

$$\Phi_\zeta(1; 1) = \zeta, \text{ for each root of } \zeta^3 - 2\zeta^2 + 1 = 0. \qquad (1.23)$$

The two positive roots are:

$$\zeta_1 = 1, \ \zeta_2 = (1 + \sqrt{5})/2. \qquad (1.24)$$

These two positive roots are known as MSS roots after Metropolis-Stein-Stein, who first introduced them (see Section 1.2.4). The relationship of this pair of MSS roots to the computer-generated inverse graphs denoted by $P2$ given in Chapter 5 and Chapter 7 is: these MSS roots are the exact creation ζ-values of the new branches of the inverse graph. Indeed, the entire ζ-evolution of the set of inverse graphs that exist for the interval $(0, 1]$ can now described as follows for all $\zeta \in [0, \infty)$:

For $\zeta \in (0, 1]$, there appears in the inverse graph the single primordial p-curve, which is the union of the upper positive branch $\Phi_\zeta((1, -1); x)$ and the conjugate branch $\Phi_\zeta((-1, -1); x)$, which join smoothly together at the extremal point $x = \zeta$ on the central line $y = 1$ of the graph. This **central right-moving** p-curve is the only curve in the inverse graph for $\zeta \in (0, 1]$. At the MSS root $\zeta_1 = 1$, this p-curve is split apart by the creation of two new branches $\Phi_\zeta((1, 1); x)$ and $\Phi_\zeta((-1, 1); x)$ that constitute the upper positive branch $\Phi_\zeta((1, 1); x)$ and the conjugate branch $\Phi_\zeta((-1, 1); x)$ of a **new**

central left-moving p-curve with its extremal point $x = \zeta$ on the central line $y = 1$. The "pushed-apart" primordial branches $\Phi_\zeta((1, -1); x)$ and $\Phi_\zeta((-1, -1); x)$ remain in the graph for all greater ζ. The picture of the final curve for all $\zeta \in [1, \infty)$ is that of two right-moving p-curves and a single left-moving p-curve that together constitute a single continuous curve. The left-moving curve evolves to $x \to -\infty$, the right-moving curve to $x \to \infty$.

Special features of the collection of $P2$ inverse graphs given in Chapter 7 that should be noted are summarized next:

1. Once a branch is created, it remains in the inverse graph for all greater values of ζ.

2. The MSS root $\zeta_2 = (1 + \sqrt{5})/2$ is not the creation value of new branches, but rather is the creation point of two new fixed points emerging out of an already existing fixed point, as shown in the computer-generated graph in the $P2$ collection in Chapter 7 labeled by $\zeta = 1.62 \approx (1 + \sqrt{5})/2$. These computer-generated graphs show the ζ-evolution of the inverse graph G_ζ^n at various specified values of ζ that are sometimes referred to as "snapshots."

3. There is no MSS root greater than 2 for $\zeta \in (0, \infty)$. This can be shown generally by considering the set of n points $\{(x, y)\} \subset \mathbb{R}^2$ obtained from the iteration of (1.19) given by

$$\Big(q_0(\zeta), q_1(\zeta)\Big), \Big(q_1(\zeta), q_2(\zeta)\Big), \ldots, \Big(q_{n-1}(\zeta), q_n(\zeta)\Big). \qquad (1.25)$$

   For $\zeta \geq 2$, each $q_1(\zeta) = \zeta \geq 2$ and each $q_i(\zeta) \leq 0, i = 2, 3, \ldots$; hence, the condition $q_n(\zeta) = 1$ cannot be fulfilled.

4. The set of n points in the plane $(x, y) \in \mathbb{R}^2$ defined by (1.25) for $0 \leq \zeta \leq 2$ define a closed path in the plane. This path is obtained by drawing a horizontal line from the starting point $(1, \zeta)$ on the parabola to the point $(q_1(\zeta), q_1(\zeta)) = (\zeta, \zeta)$ on the 45°-line, a vertical line from $(q_1(\zeta), q_1(\zeta))$ on the 45°-line to $(q_1(\zeta), q_2(\zeta))$ on the parabola, a horizontal line from $(q_1(\zeta), q_2(\zeta))$ on the parabola to $(q_2(\zeta), q_2(\zeta))$ on the 45°-line,..., a vertical line from $(q_{n-1}(\zeta), q_{n-1}(\zeta))$ on the 45°-line to $(q_{n-1}(\zeta), q_n(\zeta))$ on the parabola, where the condition that the path be closed is $(q_{n-1}(\zeta), q_n(\zeta)) = (1, \zeta)$. The path defined by these points and the corresponding x-coordinates belonging to the parabola are presented as follows for all $n \geq 2$:

   path: $(1, \zeta) \longrightarrow (\zeta, \zeta) \longrightarrow (\zeta, q_2(\zeta)) \longrightarrow (q_2(\zeta), q_2(\zeta))$
   $\longrightarrow (q_2(\zeta), q_3(\zeta)) \longrightarrow \cdots \longrightarrow (q_{n-1}(\zeta), q_{n-1}(\zeta)) \longrightarrow (q_{n-1}(\zeta), q_n(\zeta));$
   $$ \qquad (1.26)$$

   x-coordinates: $1, q_1(x), q_2(x), \ldots, q_{n-1}(x)$.

   The closed-path condition is $(q_{n-1}(\zeta), q_n(\zeta)) = (1, \zeta)$, where n is the least value such that $q_{n-1}(\zeta) = 1$. Words on two letters R and L are used to describe the right (R) and left (L) distribution of the x-coordinates relative to the central coordinate $x = 1$ of the $n - 2$ points belonging to the parabola in the path in (1.26).

5. There are four cases of $\zeta \in (0, \infty)$ to consider in (1.26):

   (a) $\zeta \in (0, 1)$: The word L^{n-1} corresponds to the path in plane $\mathbb{R}^2$ given by (1.26), but now it is impossible to impose the closed-path condition, since the successive x-coordinates on the parabola satisfy the inequalities

   $$1 > q_1(\zeta) > q_2(\zeta) > \cdots > q_{n-1}(\zeta), \ n \geq 2. \qquad (1.27)$$

   Thus, it is impossible to satisfy $q_{n-1}(\zeta) = 1$.

   (b) $\zeta = 1$: The point $(1, 1)$ already belongs to the 45°-line, that is, is a fixed point of the parabola.

   (c) $\zeta \in (1, 2)$: The simplest case occurs for $n = 3$ and has $\zeta = \zeta_2 = (1+\sqrt{5})/2$; the path contains the four points $(1, \zeta_2), (\zeta_2, \zeta_2), (\zeta_2, 1), (1, 1)$, and the x-coordinates on the parabola are $1 < \zeta_2$; hence, R is associated with this path. The golden ratio enters at the most fundamental level. The standard symbol for the golden ratio is $\phi = (1 + \sqrt{5})/2$; it has the exact value given by the continued fraction expansion containing all 1's, as expressed by

   $$\phi = 1 + \cfrac{1}{1 + \cfrac{1}{1 + \cfrac{1}{\ddots}}} . \qquad (1.28)$$

   Thus, the golden ratio satisfies the relation

   $$\phi = 1 + \frac{1}{\phi}, \qquad (1.29)$$

   which is the positive root $\phi = (1 + \sqrt{5})/2$ of the quadratic relation $\phi^2 = \phi + 1$.

   (d) $\zeta \in [2, \infty)$: Each polynomial $q_k(\zeta) \leq 0, k \geq 2$; hence, it is impossible to have a closed path.

The significance of the MSS roots of $p_{n-1}(\zeta) = q_{n-1} - 1 = 0$ is that of giving all values of ζ for which such closed paths are possible. There is a unique closed path corresponding to a word $RL^{\alpha_0 - 1} RL^{\alpha_1 - 1} \ldots RL^{\alpha_k - 1}$, $\alpha = (\alpha_0, \alpha_1, \ldots, \alpha_k) \in \mathbb{A}_n$, where $\mathbb{A}_n$ is defined to be the set of all sequences

$$\mathbb{A}_n = \{(\alpha_0, \alpha_1, \ldots, \alpha_k) \,|\, \text{each } \alpha_i \in \mathbb{P}, \ \textstyle\sum_{i=0}^k \alpha_i = n, k = 0, 1, \ldots, n\}. \quad (1.30)$$

Here $\mathbb{P}$ denotes the set of all positive integers. The set of all conjugate sequences is defined from the set of all positive sequences by

$$\overline{\mathbb{A}}_n = \{\overline{\alpha} = (-\alpha_0, -\alpha_1, \ldots, -\alpha_k) \,|\, \alpha \in \mathbb{A}_n\}. \qquad (1.31)$$

The sequence $\alpha = (1)$ gives the first such path R and corresponds to the ζ-value given by the golden ratio.

1.1.1   Words on Two Letters

Words in the two letters R and L are introduced naturally into chaos theory via the mappings $1 \mapsto R$ and $-1 \mapsto L$. This leads naturally to the following notations for the set of all words:

$$
\begin{aligned}
w(\alpha) &= RL^{\alpha_0-1}RL^{\alpha_1-1}\cdots RL^{\alpha_k-1}; \\
w(\overline{\alpha}) &= L^{\alpha_0}RL^{\alpha_1-1}\cdots RL^{\alpha_k-1}; \\
&\alpha \in \mathbb{A}_n.
\end{aligned}
\tag{1.32}
$$

Notice, then, that the conjugate word $w(\overline{\alpha})$ to $w(\alpha)$ is given by

$$
w(\overline{\alpha_0,\alpha_1,\ldots,\alpha_k}) = L^{\alpha_0}w(\alpha_1,\alpha_2,\ldots,\alpha_k), \ k \geq 1.
\tag{1.33}
$$

For example, the conjugate to RL^{α_0-1} is L^{α_0}. *It is not the interchange of R and L.*

1.1.2   A New Notation for Branches

The natural occurrence of words on the two letters R and L and their conjugates in (1.32)-(1.33) leads also to a new notation for the branches of every inverse graph, as given by

$$
\Psi_\zeta(\tau;x), \ \tau \in \mathbb{A}_n \cup \overline{\mathbb{A}}_n.
\tag{1.34}
$$

This definition is now augmented by defining a total-order relation on the set of all sequences $\tau \in \mathbb{A}_n \cup \overline{\mathbb{A}}_n$, which, in turn, will provide information about the branch functions $\Psi_\zeta(\tau;x)$ themselves.

1.1.3   A Total-Order Relation on τ Sequences

For each

$$
\tau = (\tau_0,\tau_1,\ldots,\tau_k) \in \mathbb{A}_n \cup \overline{\mathbb{A}}_n,
\tag{1.35}
$$

define the alternating sequence $A(\tau)$ with 0's adjoined by

$$
A(\tau) = (\tau_0,-\tau_1,\tau_2,\ldots,(-1)^k\tau_k,0,0, \ldots).
\tag{1.36}
$$

The zeros are adjoined to the right end of the sequence so as to give the same number of parts (counting zeros) to every such sequence. Next, for each pair of such sequences τ and τ', form the difference sequence:

$$
A(\tau) - A(\tau') = (\tau_0 - \tau_0', -\tau_1 + \tau_1', \tau_2 - \tau_2',\ldots).
\tag{1.37}
$$

Then, the parts in this difference sequence are integers, positive, zero, and negative. The following terminology defines what is meant by reverse-lexicographic order: $\tau > \tau'$, if the first nonzero term in (1.37) is positive; $\tau < \tau'$, if the first nonzero term in (1.37) is negative; $\tau = \tau'$, if all terms are equal in (1.37).

It follows from the above definition of reverse-lexicographic sequences that such sequences can also be expressed in terms of positive sequences alone and the definition of conjugate sequences. It is convenient to give this form so as to minimize the use of the somewhat cumbersome τ sequence notation. Because of the importance of this order relation, it is stated fully so as to be clear in its implementation:

$$\alpha = (\alpha_0, \alpha_1, \ldots, \alpha_k), \text{ each } \alpha_i \in \mathbb{P}, \ i = 0, 1, \ldots, k; \tag{1.38}$$
$$A(\alpha) = (\alpha_0, -\alpha_1, \alpha_2, \ldots, (-1)^k \alpha_k, 0, 0, \ldots .);$$

$\alpha > \alpha'$, if and only if the first nonzero part
of $A(\alpha) - A(\alpha')$ is positive;

$\alpha < \alpha'$, if and only if the first nonzero part $\tag{1.39}$
of $A(\alpha) = A(\alpha')$ is negative;

$\alpha = \alpha'$, if and only if all parts
of $A(\alpha) - A(\alpha')$ are equal.

The order relation for negative (conjugate) sequences can, of course, be obtained directly from the results (1.39) for positive sequences and the reflection symmetry of all inverse graphs through the $y = 1$ central line.

1.1.4 Reality of the Inverse Graph

The inverse graph G_ζ^n approach, as used in this monograph, in place of the graph $H_n(\zeta), \zeta \in (0, \infty)$ to determine properties of these two graphs, especially, the property of a unique fully deterministic subclass, shifts the focus to the following:

The subclass of computer-generated inverse graphs become the dominate objects to investigate at the outset, including a proof of their existence, of their nontriviality, and of their more important properties. (1.40)

The issue of nontriviality is especially important, as is also the classification of fixed points.

Before considering the results still required in (1.40), it is useful to give an overview of the features in question. For example, the central point $(1, 1)$ is a fixed point of each graph H_ζ^n at $\zeta = 1$, as well as G_ζ^n at $\zeta = 1$, this holding for all 2^{n-1} positive branches of G_ζ^n for $n = 1, 2, \ldots$ and the conjugate branches as well. It is already clear from definition (1.17) that:

$$\Phi_\zeta\Big((1, -1); x\Big) \text{ for } 0 \leq x \leq \zeta \text{ is real, if and only if } \sqrt{1 - \tfrac{x}{\zeta}} \geq 1 - \zeta.$$
$$\tag{1.41}$$

But there are infinitely many values for which the second $\geq$ relation is violated; for example, it is violated already at $\zeta = 1/2$ and all $0 \leq x \leq 1/2$. This already signifies that the subclass of computer-generated graphs in statement (1.40) exists for this case and is nontrivial.

Relation (1.41) already implies the general result embodied in (1.40). This is because conditions (1.41) are themselves a subclass of (1.40):

> *The important class of deterministic-computer-generated*
>
> *inverse graphs aways exists and is nontrivial.* $\hspace{2cm}$ (1.42)

It is the important features of the set of deterministic-computer generated inverse graphs that remains to be worked out. In one sense or another, it is the principal purpose of the remainder of this monograph to fill-in many of these details:

Hereafter, the notation $\Psi(\alpha; x)$ **always** denote such a deterministic-generated-inverse graph, **unless noted otherwise.** Carrying this out requires giving a great deal of supplemental information that is next developed.

1.2   Inverse Graphs Created at $\zeta = 1$

The motion of the primordial p-curve initiates the entire process of creating the inverse graph for general n: it is universal. The creation of new branches at $\zeta = 1$ can also be given in a quite nice general form for $n \geq 2$. For this reason, the description of these new branches and their dynamical fixed points is developed next. It is a good example of a class of deterministic-computer-generated inverse graphs.

The first creation event beyond the appearance of the primordial curve and its fixed point within the interval $\zeta \in (0,1]$ takes place at $\zeta = 1$. Here a family of p-curves is created simultaneously with a new central p-curve that can be described for arbitrary n as follows:

1. The primordial p-curve $C_\zeta^n\left((n) \middle| \overline{(n)}\right)$ is split at $\zeta = 1$ by the collection of n new p-curves with positive and negative branches labeled from top-to-bottom in the inverse graph G_ζ^n by the following deterministic-computer-generated $\Psi_\zeta(\alpha; x)$ functions and their conjugates:

2. The so-called **universal branches** (see Ref. [15]) are given by:

$$\Psi_\zeta((n-r+1\ 1^{r-1}); x) \text{ and } \Psi_\zeta(\overline{(n-r+1\ 1^{r-1})}; x),$$
$$r = 1, \ldots, n; \ x \in (1, \zeta_2], \hspace{2cm} (1.43)$$

where this interval is closed at ζ_2, which is the least MSS root ≥ 1 for which the interval $(1, \zeta_2]$ is closed for specified n; this MSS root is unique for each n.

3. The primordial central p-curve $C_\zeta^n\left((n) \middle| \overline{(n)}\right)$ has now been replaced by a new central p-curve $C_\zeta^n\left((1^n) \middle| \overline{(1^n)}\right)$. This p-curve is central in the inverse graph for all $x \in (1, \zeta_2]$.

4. The motion of the original fixed point $x = 2 - \frac{1}{\zeta}$ during the synchronous creation of new p-curves (1.43) is to move onto the upper branch $\Psi_\zeta((1^n); x)$ of the new central p-curves (1.43), where it remains for all $\zeta > 1$.

5. The branches constituting the original primordial p-curve have the property that all the new deterministic-computer-generated inverse in (1.43) fall between the original branches $\Psi_\zeta((n); x)$ and $\Psi_\zeta(\overline{(n)}; x)$, of the primordial p-curve branch parts; all of these branches join smoothly together at their left and right extremal points to constitute the compound p-curve that gives the space G_ζ^n for each $\zeta \in (1, \zeta_2)$. The labels of the branches are ordered by

$$(n) > (n-1\ 1) > \cdots > (2\ 1^{n-1}) > (1^n) >$$
$$\overline{(1^n)} > \overline{(2\ 1^{n-1})} > \cdots > \overline{(n)}. \tag{1.44}$$

The ordering in (1.44) coincides with that of the corresponding branches from top-to-bottom in the inverse graph. The $2n$ ordered branches constituted from these n p-curves remain in the graph for all $\zeta \in (\zeta_2, \infty)$, but are themselves further split apart by the creation of new p-curves for $\zeta > \zeta_2$. In particular, the central p-curve $C_\zeta^n\left((1^n)\,\middle|\,\overline{(1^n)}\right)$ for the interval $\zeta \in (1, \zeta_2]$ splits apart at the first MSS root greater than ζ_2.

The creation of universal inverse graphs described above is shown in the following graphs in Chapter 7 for $n = 2, 3$:

$n = 2$. The universal inverse graph is the central p-curve $C_\zeta^2\left((1\ 1)\,\middle|\,\overline{(1\ 1)}\right)$ created at $\zeta = 1$. It remains in the graph for all $\zeta > 1$. No other inverse graphs are created for $\zeta > 1$.

$n = 3$. The universal inverse graphs are created at $\zeta = 1$ and consist of the four inverse graphs labeled from top-to-bottom by the branches $(2\ 1) > (1\ 1\ 1) > \overline{(1\ 1\ 1)} > \overline{(2\ 1)}$. For $\zeta > 1$, some of the universal graphs are split apart by the creation of new inverse graphs, some are not. It is also useful to note that in terms of the notation for branches introduced in (1.34), the following relations hold for the numerical values of the branch functions at $\zeta = 2$:

$$\Psi_\zeta((3); x) \geq \Psi_\zeta((2\ 1); x) \geq \Psi_\zeta((1\ 1\ 1); x) \geq \Psi_\zeta((1\ 2); x) >$$
$$\Psi_\zeta(\overline{(1\ 2)}; x) \geq \Psi_\zeta(\overline{(1\ 1\ 1)}; x) \geq \Psi_\zeta(\overline{(2\ 1)}; x) \geq \Psi_\zeta(\overline{(3)}; x) \text{ at } \zeta = 2, \tag{1.45}$$

where the only place an equality can occur is at the common point shared by two p-curves. For $0 < \zeta < 2$, the subset of relations (1.44) holds that is obtained by striking out all branch functions not yet created for the selected value of ζ. It is also the case that relation (1.44) holds for all $\zeta \in (2, \infty)$, there being no new branch functions created for such ζ. The Ψ, of course, refer to the deterministic-computer-generated inverse graphs.

1.2.1   The Order Rule for General Branches

The order rule for $n = 3$ introduced in (1.44) extends to all branches of the general inverse graph G_ζ^n. The full graph G_ζ^n must have the following structural form at each value of the parameter ζ: It is the union over α of all real branches $G_\zeta^n(\alpha)$ and $G_\zeta^n(\overline{\alpha})$ present in the inverse graph at the given value of ζ:

$$G_\zeta^n = \bigcup_{\alpha \in \widehat{\mathbb{A}}_n(\zeta)} \left(G_\zeta^n(\alpha) \cup G_\zeta^n(\overline{\alpha}) \right). \qquad (1.46)$$

The set $\widehat{\mathbb{A}}_n(\zeta) \subseteq \mathbb{A}_n$ is defined to be the subset of $\mathbb{A}_n$ (see (1.30)) such that each branch function $\Psi_\zeta(\alpha; x)$, $\alpha \in \widehat{\mathbb{A}}_n(\zeta)$ is real, hence, the corresponding conjugate branch function is also real. The set $\widehat{\mathbb{A}}_n(\zeta)$ is yet to be determined for general n, but the form (1.46) must prevail.

The main result is that reverse-lexicographic order on sequences $\alpha \in \mathbb{A}_n$ implies the following order relation on α sequences:

Consider any two real inverse functions $\Psi_\zeta(\alpha; x)$ and $\Psi_\zeta(\alpha'; x)$ for $\alpha, \alpha' \in \mathbb{A}_n$, each of which always has a well-defined domain of definition $\mathbb{D}_\zeta(\alpha)$ and $\mathbb{D}_{\zeta'}(\alpha')$. Then, the following relations hold between the numerical values of these deterministic-computer-generated inverse functions:

$$\begin{aligned} \Psi_\zeta(\alpha; x) &> \Psi_\zeta(\alpha'; x), \text{ for } \alpha > \alpha'; \\ \Psi_\zeta(\alpha; x) &= \Psi_\zeta(\overline{\alpha}; x), \text{ for } x \text{ a unique point} \qquad (1.47) \\ &\qquad \text{on the central line } y = 1; \\ \Psi_\zeta(\overline{\alpha}; x) &< \Psi_\zeta(\overline{\alpha}'; x), \text{ for } \overline{\alpha} > \overline{\alpha}'. \end{aligned}$$

The last collection of relations for conjugate branches is implied by the other two: The labels of the conjugate branches and functions as read from the bottom upward to the central line are obtained by conjugating the sequences.

It is useful to emphasize again how the combinatorial theory of words on two letters R and L makes it appearance into chaos theory as presented in this monograph as introduced in (1.32) and given by the one-to-one correspondence:

$$\begin{aligned} \alpha &\mapsto w(\alpha) = RL^{\alpha_0-1}RL^{\alpha_1-1} \cdots RL^{\alpha_k-1}, \text{ each } \alpha \in \mathbb{A}_n, \qquad (1.48) \\ \overline{\alpha} &\mapsto w(\overline{\alpha}) = L^{\alpha_0}RL^{\alpha_1-1}RL^{\alpha_2-1} \cdots RL^{\alpha_k-1}, \text{ each } \alpha \in \mathbb{A}_n. \end{aligned}$$

The reverse-lexicographic order rule can be applied to the set of all words on two letters:

$$w(\alpha) > w(\alpha'); \quad w(\overline{\alpha}) < w(\overline{\alpha}'), \text{ if and only if } \alpha > \alpha'. \qquad (1.49)$$

The number of words in (1.47) beginning with R and L, respectively, is given by the number of solutions in positive integers of the relation: $\alpha_0 + \alpha_1 + \cdots + \alpha_k = n$. This solution set may be denoted by $\mathbb{A}_k(n)$; it has cardinality $|\mathbb{A}_k(n)| = \binom{n-1}{k}$. Thus, the cardinality of the set $\mathbb{A}_n$ is given by $|\mathbb{A}_n| = 2^{n-1}$.

Thus, all 2^n words in R and L are enumerated in (1.47) for $\alpha \in \mathbb{A}_n$, half beginning with R and half with L; those beginning with R correspond to positive sequences; those beginning with L to negative sequences. Because of the relationship of the positive sequence $\alpha = (\alpha_0, \alpha_1, \ldots, \alpha_k)$ to words given by (1.47), the degree $D(\alpha)$ of this sequence is defined as the total degree of the word polynomial:

$$D(\alpha) = D((\alpha_0, \alpha_1, \ldots, \alpha_k)) = \alpha_0 + \alpha_1 + \cdots + \alpha_k. \tag{1.50}$$

The combinatorics of words makes its appearance in a fundamental way. It is hard to imagine a more eloquent foundation on which to build the theory of the 2^n branches $\Psi_\zeta(\alpha; x)$ and $\Psi_\zeta(\overline{\alpha}; x)$.

The general form of the branch functions $\Psi_\zeta(\alpha; x)$ is implicit in the results given above. But their explicit form for arbitrary n needs also to be given. The graph G_ζ^n inverse to H_ζ^n is just the set of points obtained from the points $(x, y) \in H_\zeta^n \subset \mathbb{R}^2$ by reflection through the $45°$-line:

$$(y, x) \in G_\zeta^n, \text{ if and only if } (x, y) \in H_\zeta^n. \tag{1.51}$$

Thus, the points $(x, y) \in G_\zeta^1 \subset \mathbb{R}^2$ have x- and y-coordinates related by $x = \zeta y(2 - y)$; that is, y is given in terms of x by the square-root relations $y = 1 + \sqrt{1 - \frac{x}{\zeta}}$ and $y = 1 - \sqrt{1 - \frac{x}{\zeta}}$, subject to the conditions that all square-roots be real. Similarly, the points $(x, y) \in G_\zeta^2 \subset \mathbb{R}^2$ have x- and y-coordinates related by $x = \zeta^2 y(2 - y)\left(2 - \zeta y(2 - y)\right)$; that is, y is given in terms of x by relations (1.17), subject to the conditions that all square-roots be real. As illustrated by (1.17), it is function composition of inverse branches that defines all inverse graphs. This is just the subclass of deterministic-computer-generated inverse graphs.

The construction of the 2^n branches of the inverse functions to H_ζ^n at level n can be given in terms of the 2^{n-1} inverse branches of the inverse functions to H_ζ^{n-1} at level $n - 1$ as follows:

$$\Phi_\zeta\Big((\sigma_1, \sigma_2, \ldots, \sigma_n); x\Big) = \Phi_\zeta\Big(\sigma_1; \Phi_\zeta((\sigma_2, \sigma_3, \ldots, \sigma_n); x)\Big) \tag{1.52}$$

$$= 1 + \sigma_1 \sqrt{1 - \frac{1}{\zeta}\Phi_\zeta((\sigma_2, \sigma_3, \ldots, \sigma_n); x)}, \quad \sigma_i = 1 \text{ or } -1, i = 1, 2, \ldots, n.$$

Here, as indicated, σ_i is the i-part of σ. The iteration of this relation then gives the following explicit formula for the general inverse function in which there appears n square-root symbols in all, each one inside the preceding one:

$$\Phi_\zeta(\sigma; x) = 1 + \tag{1.53}$$

$$\sigma_1 \sqrt{1 - \frac{1}{\zeta}\left(\sqrt{1 + \sigma_2\sqrt{1 - \frac{1}{\zeta}\left(1 + \sigma_3\sqrt{1 - \cdots \sqrt{1 - \frac{1}{\zeta}\left(1 + \sigma_n\sqrt{1 - \frac{x}{\zeta}}\right)} \cdots}\right)}\right)}} \, .$$

The function $\Phi_\zeta(\sigma; x)$ is, in general, a complex function of ζ, x, even for $x \leq \zeta$. It is the enumeration of the points $(\zeta, x) \in \mathbb{R}^2$ that is one of the principal problems addressed in this monograph, as already mentioned above in the seeking the unique solution for the deterministic-computer-generated inverse graphs. It is the solution of this problem that gives all the points that constitute the branches of the inverse graph G_ζ^n for each specified $\zeta \in (0, \infty)$. It is precisely the deterministic-computer-generated inverse graphs that determine, by reflection through the $45°$-line, all points of the original real graph H_ζ^n itself. The determination of these deterministic-computer-generated solutions is nontrivial, requiring as it does a unique labeling of all such real branches and the specification of their domains.

The inverse functions $\Psi_\zeta(\alpha; x)$ for $\alpha = (\alpha_0, \alpha_1, \ldots, \alpha_k)$ and the conjugate functions $\Psi_\zeta(\overline{\alpha} : x)$ for $\overline{\alpha} = (-\alpha_0, -\alpha_1, \ldots, -\alpha_k)$, where each $\alpha_i \in \mathbb{P}$, $i = 0, 1, \ldots, k$, are defined in terms of $\Phi_\zeta(\sigma; x)$ by

$$\Psi_\zeta(\alpha; x) = \Phi_\zeta\Big(\sigma(\alpha); x\Big), \quad \text{each } \alpha \in \mathbb{A}_n;$$

$$\sigma(\alpha) = \Big(1, -1^{\alpha_0 - 1}, 1, -1^{\alpha_1 - 1}, \ldots, 1, -1^{\alpha_k - 1}\Big);$$

$$\Psi_\zeta(\overline{\alpha}; x) = \Phi_\zeta\Big(\sigma(\overline{\alpha}); x\Big), \quad \text{each } \overline{\alpha} \in \overline{\mathbb{A}}_n; \tag{1.54}$$

$$\sigma(\overline{\alpha}) = \Big(-1^{\alpha_0}, 1, -1^{\alpha_1 - 1}, \ldots, 1, -1^{\alpha_k - 1}\Big), \ k \geq 1; \ \sigma(\overline{\alpha_0}) = (-1^{\alpha_0});$$

$$\Psi_\zeta(\overline{\alpha}; x) = 2 - \Psi_\zeta(\alpha; x).$$

The $\alpha, \overline{\alpha}$ notation is preferred for the most part throughout the remainder of this monograph because it gives a clear separation into all positive α-sequences in the upper-half $y \geq 1$ of the inverse graph with the corresponding conjugate (negative) sequences always in the lower-half $y \leq 1$ of the inverse graph. Indeed, in the family of inverse graphs, the labels in the lower-half of the graph are usually assigned from the positive graphs using the reflection property through the central line $y = 1$.

1.2.2   Fixed Points

The description of the inverse graphs $\Psi_\zeta(\alpha; x), \alpha \in \mathbb{A}_n$ now continues with the very important concept of a **fixed point.** A fixed point of a graph in the (x, y)-plane is a point that belongs not only to the graph, but also to the $45°$-line. A simple fixed point for the problem at hand is the origin $(0, 0)$ of the coordinate frame. Thus, the origin $(0, 0)$ is a fixed point of the graph H_ζ^n as well as of the inverse graph G_ζ^n. This fixed point is truly fixed in that it does not change its position with changing ζ. It is verified by direct substitution into the two branches (1.6) that the point $x(\zeta) = 2 - \frac{1}{\zeta}$ is a fixed point throughout the entire ζ-evolution of the inverse graph G_ζ^1:

$$\Psi_\zeta\Big(\overline{(1)}; 2 - \frac{1}{\zeta}\Big) = 2 - \frac{1}{\zeta}, \text{ all } \zeta \in (0, 1];$$

$$\Psi_\zeta\Big((1); 2 - \frac{1}{\zeta}\Big) = 2 - \frac{1}{\zeta}, \text{ all } \zeta \in [1, \infty). \tag{1.55}$$

In verifying these relations, the standard rule that the square-root of a positive number is a positive number must be carefully observed:

$$\sqrt{1 - \frac{2}{\zeta} + \frac{1}{\zeta^2}} = \begin{cases} 1 - \frac{1}{\zeta}, & \zeta \in [1, \infty); \\ \frac{1}{\zeta} - 1, & \zeta \in (0, 1]. \end{cases} \qquad (1.56)$$

The result for the Ψ-function expressed by the second of relations (1.55) for $n = 1$ extends to all $n > 1$:

$$\Psi_\zeta\left((n); 2 - \frac{1}{\zeta}\right) = 2 - \frac{1}{\zeta}, \quad \text{all } \zeta \in [1, \infty), \ n \geq 2. \qquad (1.57)$$

This result is proved from the following composition relation and the initial condition (1.55):

$$\Psi_\zeta\left((n); 2 - \frac{1}{\zeta}\right) = \Psi_\zeta\left((1); \Psi_\zeta((n-1); 2 - \frac{1}{\zeta})\right), \ \text{all } \zeta \in [1, \infty), \ n \geq 2. \qquad (1.58)$$

Similar considerations yield:

$$\Psi_\zeta\left((1^n); 2 - \frac{1}{\zeta}\right) = 2 - \frac{1}{\zeta}, \ \text{all } \zeta \in [1, \infty), \ n \geq 2. \qquad (1.59)$$

1.2.3 The Primordial p-Curve and Its Evolution

The description of the inverse graph G_ζ^n for ζ in the interval $\zeta \in (0, 1]$ can already be given. It is convenient now to denote the primordial p-curve by the set of points in the plane $\mathbb{R}^2$ defined as follows:

$$C_\zeta^n\left((n) \,\middle|\, \overline{(n)}\right) = G_\zeta^n((n)) \cup G_\zeta^n(\overline{(n)}) \qquad (1.60)$$

$$= \left\{\left(x, \Psi_\zeta((n); x)\right) \,\middle|\, x \in [0, \zeta]\right\} \bigcup \left\{\left(x, \Psi_\zeta(\overline{(n)}; x)\right) \,\middle|\, x \in [0, \zeta]\right\}.$$

In terms of this notation, a single p-curve $C_\zeta^n\left((n) \,\middle|\, \overline{(n)}\right)$ is present in the graph G_ζ^n for ζ in the interval $\zeta \in (0, 1]$. The ζ-evolution of this primordial curve p-curve goes as follows: The curve emerges at $x = 0$ out of the vertical line-segment $y \in [0, 2]$ at $x = 0$; for large n, it is almost a square curve of the shape pictured by

primordial p-curve

$\Phi_\zeta(n; x)$: upper branch $\qquad (1.61)$

$\Phi_\zeta(\overline{n}; x)$: lower branch

The upper and lower horizontal lines and square-corners defining the square shape are always somewhat curved, more so for small n. As ζ increases past

0, the primordial p-curve remains centered above and below the central line $y = 1$ and moves toward the right in the graph, becoming more and more square as the common central point of the p-curve approaches the point $(1, 1)$ of the graph. As mentioned above, this dynamical p-curve is the only curve in the graph for $\zeta \in (0, 1]$. But throughout this period of ζ-evolution, there is also present the dynamical fixed point $x(\zeta) = 2 - \frac{1}{\zeta}$; it seems to emerge out of the fixed point $(0, 0)$ at the origin, but a glance at any computer-generated graph for negative x shows that it is already present to the left of the origin. For all positive ζ, it moves smoothly along the 45°-line, belonging at the same time to the conjugate function $\Psi_\zeta(\overline{n}; x)$, until the point $(1, 1)$ of the graph is reached. The primordial p-curve and its fixed point $x(\zeta) = 2 - \frac{1}{\zeta}$ constitute the full inverse graph for all $\zeta \in (0, 1]$.

The p-curves that constitute the deterministic-computer-generated graph at each value of the parameter ζ may be considered as defining an abstract space. In this viewpoint, it is only the set of p-curves themselves that enter. The coordinate system itself and as well as the 45°-line are not part of the abstract space. They only serve to help visualize the space, which is fully composed of the p-curves themselves. It is also useful to speak of the objects in the abstract space. For example, all **dynamical fixed points** of the inverse graph may be taken as **a set of objects** contained in the abstract space defined by the shape of the curve present in G^n_ζ for each $\zeta \in (0, \infty)$. To repeat: *The entire abstract space is just the set of p-curves constituting the inverse graph at each value of ζ; it is often the case that the dynamical fixed points are chosen as the objects in this abstract deterministic-computer-generated space, but this is not required.*

The 45°-line has another important visualization role: The formation of fixed points is preceded by a branch of the inverse graph becoming tangent to the 45°-line. Thus, as a left-moving or right-moving p-curve approaches the 45°-line, its extremal point meets that line, crosses it, and becomes exactly tangent to it. For the primordial p-curve, the tangency occurs at exactly one ζ-value in the interval $\zeta \in (0, 1]$; namely, $\zeta = 1/2$ on the conjugate inverse graph as expressed by the following derivative relations:

$$\frac{d}{dx}\Psi_\zeta(\overline{1}; x)\Big|_{x=2-\frac{1}{\zeta}} = \frac{1}{2(1 - \zeta)} = 1, \text{ at } \zeta = 1/2;$$

$$\frac{d}{dx}\Psi_\zeta(\overline{n}; x)\Big|_{x=2-\frac{1}{\zeta}} \tag{1.62}$$

$$= \frac{1}{2\zeta\sqrt{1 - \frac{1}{\zeta}\Psi_\zeta\left(\overline{n-1}; 2 - \frac{1}{\zeta}\right)}} \frac{d}{dx}\Psi_\zeta(\overline{n-1}; x)\Big|_{x=2-\frac{1}{\zeta}}$$

$$= \frac{1}{2(1 - \zeta)}\frac{d}{dx}\Psi_\zeta(\overline{n-1}; x)\Big|_{x=2-\frac{1}{\zeta}} = 1, \text{ at } \zeta = 1/2.$$

The last step follows from the induction hypothesis at level $n - 1$, and the full induction proof from the validity of the first relation at level $n = 1$.

It is useful here to give the inverse graph to which the dynamical fixed point $x(\zeta) = 2 - \frac{1}{\zeta}$ belongs as its ζ evolution increases beyond the interval $(0,1]$ into the interval $\zeta \in (1, \infty)$: It is always the case for arbitrary $n \geq 2$ that new branches are created at $\zeta = 1$. In particular, the function $\Psi_\zeta((1^n); x)$ becomes real at $\zeta = 1$, and for the subclass of deterministic-computer-generate inverse graph remains real for all $\zeta \in (1, \infty)$, and has a nonempty domain of definition. Indeed, the p-curve $G_\zeta^n(1^n) \cup G_\zeta^n(-1^n)$ is central for $\zeta \in (1, \zeta_2]$, where ζ_2 is the first MSS root greater than 1, which is $(1 + \sqrt{2})/2$ for $n = 2$. At the point $(1, 1)$ of the graph, the fixed point $x(\zeta) = 2 - \frac{1}{\zeta}$, previously on the conjugate branch $G_\zeta^n(\overline{n})$ for $\zeta \in (0, 1]$, moves smoothly onto the central positive branch $G_\zeta^n(1^n)$, where it remains for all $\zeta \in (1, \infty)$; it remains on this graph even after this branch is no longer central: *The original primordial fixed point stays on the branch $G_\zeta^n(1^n)$ for all $\zeta > 1$.*

The above is a quite complete description of the ζ-evolution of every inverse graph G_ζ^n, $n \geq 1$, for the special interval $\zeta \in (0, 1]$. The description of the ζ-pathway of the creation of new p-curves and fixed points for $\zeta > 1$ is an intricate task, requiring many new concepts and their development. This introductory chapter continues with this.

1.2.4   MSS Polynomials and Roots

The real positive roots of some general polynomials known as MSS polynomials, named after the authors N. Metropolis, *et al* [3], who first introduced them, have a definitive role in the description of the ζ-values where new p-curves are created, as described for $n = 2$ in the above. The MSS polynomials originally were introduced from the repeated iteration of the parabolic map $p_\zeta^1(x) = \zeta\, x(2 - x)$, starting at $x = 1$ with the (maximal) value $p_\zeta^1(1) = \zeta$ and requiring, after n such iterations, a return to the original starting point $x = 1$; that is, $p_\zeta^n(1) = \zeta$. This iteration leads to the MSS polynomials, denoted $q_n(\zeta)$, that satisfy the following nonlinear recurrence relation of the same form as the iteration of the basic parabola itself:

$$q_{n+1}(\zeta) = \zeta\, q_n(\zeta)(2 - q_n(\zeta)), \quad n = 0, 1, 2, \ldots; \quad q_0(\zeta) = 1. \tag{1.63}$$

This definition then leads immediately to the general form:

$$q_n(\zeta) = p_\zeta^n(1), \quad n = 1, 2, \ldots . \tag{1.64}$$

It is sometimes useful to use in place of $q_n(\zeta)$ the polynomials $p_n(\zeta)$ defined by

$$p_n(\zeta) = 1 - q_n(\zeta), n = 0, 1, 2, \ldots . \tag{1.65}$$

These polynomials are also called MSS polynomials; they are fully defined by the following nonlinear recurrence relation obtained from (1.63):

$$p_n(\zeta) = \zeta\Big(p_{n-1}(\zeta)\Big)^2 - \zeta + 1, \quad n = 1, 2, \ldots; p_0(\zeta) = 0. \tag{1.66}$$

The MSS polynomial $p_n(\zeta)$ is of degree 2^{n-1} with leading coefficient 1 (a monic polynomial) for $n > 1$. The first six are:

$$p_1(\zeta) = -\zeta + 1,$$

$$p_2(\zeta) = \zeta^3 - 2\zeta^2 + 1,$$

$$p_3(\zeta) = \zeta^7 - 4\zeta^6 + 4\zeta^5 + 2\zeta^4 - 4\zeta^3 + 1,$$

$$p_4(\zeta) = \zeta^{15} - 8\zeta^{14} + 24\zeta^{13} - 28\zeta^{12} - 8\zeta^{11} + 48\zeta^{10}$$
$$- 28\zeta^9 - 14\zeta^8 + 8\zeta^7 + 8\zeta^6 + 4\zeta^5 - 8\zeta^4 + 1,$$

$$p_5(\zeta) = \zeta^{31} - 16\zeta^{30} + 112\zeta^{29} - 440\zeta^{28} + 1008\zeta^{27} - 1120\zeta^{26}$$
$$- 424\zeta^{25} + 3172\zeta^{24} - 3728\zeta^{23} + 16\zeta^{22} + 3800\zeta^{21}$$
$$- 2608\zeta^{20} - 816\zeta^{19} + 816\zeta^{18} + 900\zeta^{17} - 158\zeta^{16}$$
$$- 1168\zeta^{15} + 512\zeta^{14} + 296\zeta^{13} - 80\zeta^{12} - 16\zeta^{11} - 120\zeta^{10}$$
$$+ 36\zeta^9 + 16\zeta^8 + 16\zeta^7 + 8\zeta^6 - 16\zeta^5 + 1,$$

$$p_6(\zeta) = \zeta^{63} - 32\zeta^{62} + 480\zeta^{61} - 4464\zeta^{60} + 26640\zeta^{59} \qquad (1.67)$$
$$- 133056\zeta^{58} + 454384\zeta^{57} - 1118008\zeta^{56} + 1797728\zeta^{55}$$
$$- 1054944\zeta^{54} - 3219728\zeta^{53} + 10501920\zeta^{52} - 13522208\zeta^{51}$$
$$+ 1792672\zeta^{50} + 22935832\zeta^{49} - 36561980\zeta^{48} + 14460192\zeta^{47}$$
$$+ 28883392\zeta^{46} - 44337552\zeta^{45} + 12496544\zeta^{44} + 22471648\zeta^{43}$$
$$- 16717040\zeta^{42} - 6575528\zeta^{41} + 2982496\zeta^{40} + 15210400\zeta^{39}$$
$$- 9370768\zeta^{38} - 11209568\zeta^{37} + 12256192\zeta^{36} + 1348048\zeta^{35}$$
$$- 4074704\zeta^{34} - 1663740\zeta^{33} + 1088194\zeta^{32} + 2475808\zeta^{31}$$
$$- 1490432\zeta^{30} - 530608\zeta^{29} + 290528\zeta^{28} - 19776\zeta^{27}$$
$$+ 329664\zeta^{26} - 140792\zeta^{25} - 93216\zeta^{24} - 11520x^{23} + 17904\zeta^{22}$$
$$+ 60512\zeta^{21} - 24992\zeta^{20} - 12176\zeta^{19} + 1416^{18} - 316\zeta^{17}$$
$$+ 2592\zeta^{16} + 384\zeta^{15} + 336\zeta^{14} - 608\zeta^{13} - 288\zeta^{12} + 16\zeta^{11}$$
$$+ 72\zeta^{10} + 32\zeta^9 + 32\zeta^8 + 16\zeta^7 - 32\zeta^6 + 1.$$

Several important properties of MSS roots, which, by definition, are **the real positive roots of an MSS polynomial,** follow:

1. Each MSS polynomial has $\zeta = 1$ as a root; that is, the sum of the coefficients is 0.

2. Each MSS root of $p_n(\zeta) = -q_n(\zeta) + 1 = 0$ belongs to the interval $[1,2)$ for all $n = 1, 2, 3, \ldots$.

3. Special values of the MSS polynomial $p_n(\zeta)$ are $p_n(0) = p_n(2) = 1$, $n \geq 2$.

4. The number of MSS roots of $p_n(\zeta)$ is given by $\sum_{m|n} |\mathbb{L}_m|$, where $\mathbb{L}_m$ denotes the set of lexical sequences of degree $m-1$ defined in Sect. 1.2.5 just below, and $m|n$ denotes that m divides n, which includes both 1 and n, and $|\mathbb{L}_1| = 1$.

That $\zeta = 1$ is a root follows by induction from the recurrence relation (1.65), as also does $p_n(0) = p_n(2) = 1$, for $n \geq 2$. Similarly, the recurrence relation (1.65) shows that the only common root shared by MSS polynomials is $\zeta = 1$.

1.2.5 Lexical Sequences

The concept of a lexical sequence is very important for the enumeration of ζ-intervals defined by certain MSS roots at which new p-curves are created. The definition of a lexical sequence is based on the reverse-lexicographic order relation introduced above. Let $\lambda = (\lambda_0, \lambda_1, \ldots, \lambda_k)$ denote a sequence of length $k + 1$. Then, the sequence

$$\lambda = (\lambda_0, \lambda_1, \ldots, \lambda_k) \text{ is lexical, if and only if}$$
$$\lambda > (\lambda_i, \lambda_{i+1}, \ldots, \lambda_k), \quad \text{each } i = 1, 2, \ldots, k. \tag{1.68}$$

The sequence $(\lambda_i, \lambda_{i+1}, \ldots, \lambda_k)$ is called a **right subsequence** of λ:

A positive sequence λ of length ≥ 2 is lexical if and only if it is greater than each of its right subsequences.

This definition holds for all positive sequences of length greater than 1. It is convenient, however, to define all sequences (λ_0), $\lambda_0 = 1, 2, 3, \ldots$, of length 1 to be lexical. These lexical sequences correspond to the words $(1) \mapsto R, (2) \mapsto RL, \ldots, (\alpha_0) \mapsto RL^{\alpha_0 - 1}, \ldots$. Indeed, the sequence (0) is also included among the lexical sequences; it is of length 0 and corresponds to the empty word (no word) in R and L. A sequence that is not lexical is called nonlexical. Thus, the set of positive sequences of arbitrary length is partitioned into lexical and nonlexical sequences; the lexical sequence (0) is adjoined to represent the empty word.

The following three sets of lexical sequences have an important role in explaining various features of the inverse graph:

$$\begin{aligned}
\mathbb{L}_n &= \{\lambda = (\lambda_0, \lambda_1, \ldots, \lambda_k) \mid \lambda \text{ is lexical}, 1 + D(\lambda) = n\}; \\
\mathbb{M}_n &= \mathbb{L}_1 \cup \mathbb{L}_2 \cup \cdots \cup \mathbb{L}_n; \\
\mathbb{K}_n &= \cup_{d|n} \mathbb{L}_d,
\end{aligned} \tag{1.69}$$

where $d|n$ denotes that the positive integer d divides the positive integer n. The cardinality of the lexical sequences can be given quite explicitly; it is presented here without proof, which can be found in Brucks ([39]):

$$|\mathbb{L}_n| = \frac{1}{n}(2^{n-1} - e_n);$$

$$e_n = \begin{cases} 1, & \text{for } n \text{ prime}, \quad n \geq 3, \\ 0, & \text{for } n = 2^k, \ k = 0, 1, 2, \ldots, \\ \displaystyle\sum_{d|n,\, d>1 \text{ and odd}} \frac{n}{d} |\mathbb{L}_d|, & \text{otherwise.} \end{cases} \tag{1.70}$$

For all $n \geq 3$ and nonprime and not a power of 2, this formula for $|\mathbb{L}_n|$ is recursive in structure. For n a prime number, it is one of Fermat's theorems. See relations (2.2)-(2.3) below for a short list of lexical sequences.

Lexical sequences enter into the properties of inverse graphs is several important ways: (i) Enumeration of the positive real roots of the MSS polynomials give exactly the creation points of all branches of the general inverse graph; (ii) definition of *central sequences,* which have a major role in partitioning the branches of the inverse graph into cycle classes, and (iii) in the description of the ζ-evolution of the entire inverse graph in terms of the central branches.

1.3   Preview of the Full ζ-Evolution

The results given above already give the broad picture of the ζ-evolution of the inverse graph G_ζ^n. The main interest is, of course, is the collection of branches $G_\zeta^n(\alpha)$, $\alpha \in \widehat{\mathbb{A}}_n(\zeta) \subset \mathbb{A}_n$ that are present for each value of ζ. As noted earlier, the proof of the existence of unique deterministic-computer-generated inverse graph and its properties are now the major focus of this monograph. But even with this property fully in place the various α labeling the branches of G_ζ^n come into the real domain at different critical values of ζ that depend on the particular α; these creation values of ζ for the various new branches are always an MSS root of some MSS polynomial $p_m(\zeta), 1 \leq m < n$, still to be determined. For the more detailed picture, the ζ-intervals for which the same constituent branches $\alpha \in \widehat{\mathbb{A}}_n(\zeta) \subset \mathbb{A}_n$ persist must be determined, as well as the ζ-values at which all the remaining branches $\alpha \in \mathbb{A}_n(\zeta), \alpha \notin \widehat{\mathbb{A}}_n(\zeta)$ are created. The determination of the creation ζ-value of each branch of G_ζ^n is a quite difficult task.

The order relations (1.38)-(1.39) hold for all $\alpha, \alpha' \in \widehat{\mathbb{A}}_n(\zeta) \subset \mathbb{A}_n$, and for all $\zeta > 0$. This feature is illustrated many times in the inverse graphs presented in Chapters 5 and 7. It implies that new **positive** branches of the inverse graph not present for a particular value of ζ must be created at greater ζ-values and at y-levels above the central $y = 1$ level. This already foretells the intricate manner in which the ζ-evolution of the graph must unfold in order to create all of its 2^n branches. The relation of the reflection symmetry between positive α sequences and their conjugates in simplifying the picture cannot be over emphasized.

The implementation of the above features into the labeling of the branches of the inverse graph G_ζ^n at each value of ζ requires a **covering** of the interval $\zeta \in (0, 2]$ by the subintervals that give the ζ-values for **central branches.** In this respect, it is important to emphasize again that, once a branch $G_\zeta^n(\alpha)$ has been created at a particular ζ-value, against the deterministic-computer-generated background, it remains in the inverse graph G_ζ^n for all greater ζ; similarly for the conjugate branch. Branches and their labels are invariant objects in the inverse graph in the sense that they remain in the inverse graph with the same labels after they have been created, even though their dynamical motions changes their shapes. Not all p-curves can split apart — those for which the branch labels are adjacent labels in the full ordered set $\{(n) \,|\, (1 \; n - 1)\} = \{(n) > (n - 1 \; 1) > \cdots > (1 \; n - 1 \; 1)\}$ cannot split; but all p-curves with branches labeled by non-adjacent labels must split to accommodate the creation of all 2^n branches in the full ζ-evolution of the inverse graph. It is this dynamical picture of evolving labeled branches and p-curves that must be pieced together smoothly at each value of ζ to obtain the composite, smooth features of the ζ-evolution of the deterministic-computer-generated inverse graph.

1.4 The Baseline

The concept of a baseline of central labels $\mathbf{B}_n$ is introduced to help capture the complexity of the ζ-evolution described above. A baseline of central labels $\mathbf{B}_n$ is a collection of b_n disjoint subintervals $\zeta \in (\zeta_t, \zeta_{t+1}], t = 0, 1, \ldots, b_n - 1$, that **cover** the interval $\zeta \in (0, \infty) = (0, 2] \cup (2, \infty)$; hence, have the property:

$$(0, \infty) = \left(\cup_{t=0}^{b_n-1} (\zeta_t, \zeta_{t+1}] \right) \cup (2, \infty). \tag{1.71}$$

This baseline of central labels can be presented by the picture:

Baseline $\mathbf{B}_n$ of Central Labels

$$
\begin{array}{c}
\underset{\substack{\zeta_0 = 0}}{|} \overset{(n)}{\underset{\substack{\zeta_1 = 1}}{|}} \overset{(1^n)}{\underset{\substack{\zeta_2}}{|}} \overset{\cdots}{\underset{\substack{\cdots}}{}} \overset{|(1\ n-2\ 1)|}{\underset{\substack{\zeta_{b_n-2}}}{}} \overset{(1\ n-1)}{\underset{\substack{\zeta_{b_n-1}}}{|}} \overset{\cdots}{\underset{\substack{\zeta_{b_n} = 2}}{|}}
\end{array} \tag{1.72}
$$

The numbers $b_0 = 0$ and $b_n = 2$ label the limits of the closed interval $[0, 2]$, while all other labels are those of an MSS root, which belongs to the set of all positive roots of an MSS polynomial. The set of MSS roots that enter into the definition of the baseline $\mathbf{B}_n$ for each $n \geq 1$ is yet to be given. This set of roots, which determines the limits of the intervals in this baseline, is denoted by the ordered labels.

$$
\begin{aligned}
\mathbb{C}_n &= \{c_n(t) \,|\, t = 0, 1, \ldots, b_n - 1\}^{ord} \\
&= \{(n) > (1^n) > \cdots > (1\ n-2\ 1) > (1\ n-1)\}^{ord}. \tag{1.73}
\end{aligned}
$$

These labels are those of the positive branches $\Psi_\zeta(c_n(t); x), t = 0, 1, \ldots, b_n - 1$ that are central in the respective intervals:

$$(0, \zeta_1], (\zeta_1, \zeta_2], \ldots, (\zeta_{b_n-2}, \zeta_{b_n-1}], (\zeta_{b_n-1}, 2), \text{ where } \zeta_{b_n} = 2. \tag{1.74}$$

A collection of branches created at the same MSS root is called **synchronous** (or synchronous p-curves). It is very important here to note again that the only ζ-values where any branch $\Psi_\zeta(\alpha; x)$ can be created is at an MSS root defining the baseline $\mathbf{B}_n$; all are created synchronously with the central branch; and they are dispersed by some rules into smaller collections that fall between already-created branches. This phenomenon is well-represented above: The creation process begins with the primordial branch (n) for the interval $\zeta \in (0, 1]$; continues with the creation at $\zeta = \zeta_1 = 1$ of the $n - 1$ new synchronous branches $(n-1\ 1) > (n-2\ 1\ 1) > \cdots > (1\ 1 \cdots 1) = (1^n)$; continues with the creation of a collection of synchronous branches at the next MSS root greater than 1, which is $(1 + \sqrt{5})/2$

for $n = 2; \cdots$; continues with the final creation of the last collection of synchronous branches at $\zeta = \zeta_{b_n-1}$. All 2^{n-1} branches with labels in the ordered set $\{(n) \,|\, (1 \ n - 1)\} = \{(n) > (n - 1 \ 1) > \cdots > (1 \ n - 1)\}$ now appear in the inverse graph. Each of these successive creations of branch functions includes the central sequence $c_n(t)$ as the least synchronous label in its collection; each p-curve

$$C_\zeta^n \left(c_n(t) \,\middle|\, \overline{c_n(t)} \right), \zeta \in (\zeta_t, \zeta_{t+1}], \ t = 0, 1, \ldots, b_n-1, \tag{1.75}$$

is central for all $\zeta \in (\zeta_t, \zeta_{t+1})$. At issue is the y-level at which each new synchronous branch created at the same MSS root is to be placed. This problem is present at each value of $n \geq 2$ (see (1.45)), where the general result is now presented as

$$\Psi_\zeta((n); x) \geq \Psi_\zeta((n - 1 \ 1); x) \geq \cdots \geq \Psi_\zeta((1^n); x)$$
$$\geq \Psi_\zeta(\overline{(1^n)}; x) \geq \Psi_\zeta(\overline{(n - 1 \ 1)}; x) \geq \cdots \geq \Psi_\zeta(\overline{(n)}; x). \tag{1.76}$$

There is one over-riding rule that is never violated, which is the

Order Rule: *The y-levels occupied by branches at every value of ζ are labeled by α-sequences in $\mathbb{A}_n$ from a greatest sequence to a least sequence as read from top-to-bottom.*

This rule, however, is **not** sufficient for the column placement of a sequence for general n. The examples given by (5.35)-(5.40) in Chapter 5 give the correct column placement of all sequences for $n = 1 - 6$. The baseline $\mathbf{B}_n$ pictured in (1.72) serves as the platform for an information table erected above it and called the **Table of Creation Sequences** $\mathbb{T}_n$. *It is the column placement of sequences in $\mathbb{T}_n$ that must be understood and explained.*

The notation for α-sequences given by $\alpha = (\alpha_0, \alpha_1, \ldots, \alpha_k)$ for $k = 0$ becomes (α_0), but it is often convenient to denote it by the numerical value α_0. This is ignored for the most part, since it is usually clear from the context what is intended. But it is quite important to point out that the picture (1.72) for the baseline intervals is highly symbolic, since the lengths of these intervals are shown more or less as equal for display purposes. But, in fact, the lengths of these baseline intervals past $\zeta = 1$ and toward the right become extraordinarily small. This is shown dramatically in the graphs of MSS roots in Chapter 6 as they pile-up as ζ approaches 2.

The general format of table $\mathbb{T}_n$ can be set forth, leaving aside for the moment how this format is to be filled-in with explicit sequences from $\mathbb{A}_n$: Each table has 2^{n-1} rows and a number of columns equal to the number $|\mathbb{C}_n| = b_n$ of central sequences. Each of the b_n columns contains exactly **one sequence in each row.** Then, column n is given by

$$Col_t^{(n)} = \{\text{collection of labels created synchronously}$$
$$\text{with the central sequence } c_n(t)\}. \tag{1.77}$$

The central label $c_n(t)$ of the upper branch is placed at the bottom of each column, since it is the least sequence that can occur in that column.

One of the principal goals of this monograph is to give the rules of construction for the Creation Table $\mathbb{T}_n$ for all n. For this, it is useful first to have a more detailed description and construction of the baseline $\mathbf{B}_n$, especially, of its central sequences $c_n(t)$. This is a good starting point, although many features of the general inverse graph G_ζ^n are still left out. These include topics such as the characterizations of p-curves as left-moving or right-moving for increasing ζ, of the the dynamical domains of definition of branches, and of the creation of the dynamical fixed points, their bifurcations, and the curves on which they permanently reside. Each of these topics is developed in detail in the chapters that follow.

There are also Chapters 5, 6, and 7 in which a large number of computer-generated inverse graphs are presented that verify numerically the theory. Without the close coordination of theory development with actual visualization of events in these graphs, this monograph would not have been possible, as already acknowledged in the Preface.

This Chapter is concluded with a section detailing the vocabulary, concepts, and their symbolization for the purpose of having this information accessible in one place.

1.5   Vocabulary, Symbol Definitions, and Explanations

Math Symbols

$\mathbb{R}$	real numbers
$\mathbb{C}$	complex numbers
$\mathbb{P}$	positive numbers
$\mathbb{Z}$	integers
$\mathbb{N}$	nonnegative integers
$\mathbb{R}^n$	Cartesian n-space
$\mathbb{C}^n$	complex n-space
$\mathbb{E}^n$	Euclidean n-space
$\times$	ordinary multiplication in split product, direct product
$\delta_{i,j}$	the Kronecker delta for integers i,j
$\delta_{A,B}$, $\delta(A,B)$	the Kronecker delta for sets A and B
$\lceil x \rceil$	least integer $\geq x$
$\lfloor x \rfloor$	greatest integer $\leq x$
$\{\tau\,|\,\tau'\}$	set of all adjacent sequences from $\tau \geq \tau'$
$\alpha, \beta, \gamma \ldots$	sequences in $\mathbb{A}_n$
ϕ	empty sequence
X^m	object X repeated m times; $X^\phi = \phi$
σ_i	i-th part of a sequence σ
$(\,)\,;(\,]\,;[\,)\,;[\,]$	real number intervals
$\sqrt{a}$	a positive number that includes the square root of a negative number squared
$\Psi_\zeta(\alpha; x)$	real-valued functions of branches
$\Phi_\zeta(\sigma; x)$	real-valued functions of branches
$\mathbf{B}_n$	a baseline of central sequences
$Col_t^{(n)}$, col_i	columns in baseline

The following are terms and symbols applied to sequences within the text at the needed places:

length, degree, right subsequence, left subsequence, central, lexical, order relation, harmonic, fundamental, primitive, maximal lexical, zero sequence, boundary, adjacent, conjugate sequence, branch, p-curves, golden ratio, synchronous branches, upper branch, lower branch, right-moving, left-moving combinatorics, words on two letters, compositions, concatenation, inverse graphs, fixed points, tangency point, upper half, lower half, curves, shape of a curve, domain of definition of a branch, bifurcations,baseline intervals, composition of sequence, cycle classes, MSS polynomials.

Special Subsets

$$\{\tau\,|\,\tau'\} = \begin{array}{l}\text{set of all ordered adjacent sequences from } \tau \text{ to } \tau'; \\ \text{single sequence } \tau \text{ for } \tau = \tau';\end{array}$$

$$\lfloor\tau\,|\,\tau'\rfloor = \begin{array}{l}\text{set of all ordered adjacent sequences} \\ \text{less than } \tau \text{ and greater than } \tau'; \\ \text{empty set for } \tau \text{ adjacent to } \tau; \text{ undefined for } \tau = \tau';\end{array} \qquad (1.78)$$

$$\{\tau\,|\,\tau'\rfloor = \begin{array}{l}\text{set of all ordered adjacent sequences from } \tau \text{ to less than } \tau'; \\ \text{empty set for } \tau = \tau'; \text{ undefined for } \tau = \tau'.\end{array}$$

Chapter 2

Recursive Construction

The purpose of this Chapter 2 is to fill in the details of the full recursive construction of creation table $\mathbb{T}_n$ from creation table $\mathbb{T}_{n-1}$, thereby giving the information needed to show that each such creation table $\mathbb{T}_n$ is a unique deterministic-computer-generated set of Ψ functions. It is useful to introduce a special notation for a deterministic-computer-generated set of inverse functions, although it has not yet been shown to exist:

The notation $\mathbb{A}_n^*$ denotes a set of α sequences with the property:

$$\mathbb{A}_n^* \cup \overline{\mathbb{A}}_n^* = \mathbb{A}_n \cup \overline{\mathbb{A}}_n. \tag{2.1}$$

This property is the assertion that no sequences in Table $\mathbb{T}_n$ and the set of conjugates are left out. Thus, the notation $\mathbb{A}_n^*$ always refers unambiguously to a unique set of deterministic-computer-generated inverse graphs. These are the inverse graphs presented in Chapter 5 and Chapter 7. This is to be contrasted with the much, much more complicated set of inverse graphs given by relations (1.52)-(1.53), which is an admixture of regions of complex numbers and real numbers; indeed, $\mathbb{A}_n^*$ is exactly what remains when the complex number regions are exactly eliminated. There is little motivation for studying the more complicated inverse graphs until the simplest case $\mathbb{A}_n^*$ is understood. The remainder of this monograph addresses this problem, despite the fact that the direction taken often seems to suggest otherwise.

It is the case that the set $\mathbb{A}_n^*$ must always exist. This is true because it exists for $n = 2$, as shown above for relations (1.17), and this is a subclass for general n. The issue is to exhibit as many properties as possible of the set $\mathbb{A}_n^*$ to show its richness of structure and that it is a fully deterministic-computer-generated structure.

This begins with the derivation of the general central sequence $c_n(t)$ that gives the least label in the column $Col_t^{(n)}$ of baseline $\mathbf{B}_n$ of $\mathbb{T}_n$, where the central sequences and their columns are enumerated by $t = 0, 1, \ldots, b_n - 1$. Table $\mathbb{T}_n$ then contains in $Col_t^{(n)}$ the ordered set of labels of the new branches of the inverse graph G_ζ^n created at the MSS root ζ_t that designates the left endpoint of the interval $(\zeta_t, \zeta_{t+1}]$ that defines $Col_t^{(n)}$. The problem is to find the unique labels of these central sequences.

2.1 Construction of the Baseline $\mathbf{B}_n$

The definition of the baseline $\mathbf{B}_n$ presented in Sect. 1.4 (see (1.72)) must now be augmented with the full rule for determining each central sequence $c_n(t)$. Once the baseline $\mathbf{B}_n$ is fully determined, the ordered labels that go into each $Col_t^{(n)}$ can be considered.

Let $\lambda = (\lambda_0, \lambda_1, \dots, \lambda_k) \in \mathbb{L}_d$ denote a lexical sequence in the set $\mathbb{L}_d$, $d = 1, 2, \dots$ (see Sect. 1.2.5 for the definition of lexical sequences). In particular, a lexical sequences $\lambda \in \mathbb{L}_d$ of length $k + 1$ aways satisfies $D(\lambda) = \lambda_0 + \lambda_1 + \cdots + \lambda_k = d - 1$ for each positive integer $d \geq 2$; hence,

$$\lambda = (\lambda_0, \lambda_1, \dots, \lambda_k) \in \mathbb{L}_d \text{ implies } 1 + D(\lambda) = k. \qquad (2.2)$$

In addition to this degree requirement, each sequence $\gamma \in \mathbb{L}_d, n \geq 2$, must satisfy the order conditions that each of its right subsequences is less than the sequence γ itself; furthermore, by definition, all sequences $(0), (1), (2), \dots$, are taken to be lexical. This leads to the following sets of lexical sequences:

Examples. Lexical sequences $\mathbb{L}_d$, $d = 1, 2, \dots, 7$:

$$\mathbb{L}_1 = \{(0)\}, \mathbb{L}_2 = \{(1)\}, \mathbb{L}_3 = \{(2)\},$$

$$\mathbb{L}_4 = \{(3), (2\ 1)\}, \mathbb{L}_5 = \{(4), (3\ 1), (2\ 1\ 1)\},$$

$$\mathbb{L}_6 = \{(5), (4\ 1), (3\ 1\ 1), (3\ 2), (2\ 1\ 1\ 1)\}, \qquad (2.3)$$

$$\mathbb{L}_7 = \{(6), (5\ 1), (4\ 1\ 1), (4\ 2), (3\ 1\ 2), (3\ 1\ 1\ 1),$$

$$(3\ 2\ 1), (2\ 1\ 2\ 1), (2\ 1\ 1\ 1\ 1)\}.$$

These are the complete sets of lexical sequences of the indicated degree; they have the following cardinalities:

$$|\mathbb{L}_1| = 1, |\mathbb{L}_2| = 1, |\mathbb{L}_3| = 1, |\mathbb{L}_4| = 2, |\mathbb{L}_5| = 3, |\mathbb{L}_6| = 5, |\mathbb{L}_7| = 9. \qquad (2.4)$$

The baseline $\mathbf{B}_n$ introduced in the picture (1.72) has an MSS root denoted ζ_t at the left endpoint of the interval $(\zeta_t, \zeta_{t+1}], t = 1, 2, \dots, b_n - 1$; it is the creation ζ-value of the branches in the inverse graph G_ζ^n that are labeled by the sequences in the set $Col_t^{(n)}$. (The numerical value of each MSS root must be known to high precision to calculate the inverse graphs of G_ζ^n, which are presented in Chapter 7 under the notation Pn.)

2.1.1 Properties of MSS Roots

A notation for the set of all MSS roots is next introduced so as to be able to refer unambiguously to their various properties. An MSS root is always a **positive number:**

$$\begin{aligned}
\text{MSS}_n &= \text{set of all positive roots of the MSS polynomial } p_n(\zeta); \\
\text{MSS}_n &= \{\zeta_m^{(n)} \mid m \text{ divides } n\}; \qquad (2.5) \\
|\text{MSS}_n| &= \text{number of MSS roots} = \text{number of divisors of } n.
\end{aligned}$$

The first few of these cardinalities are given by

$$\begin{aligned} |\text{MSS}_1| &= 1, |\text{MSS}_1| = 2, |\text{MSS}_3| = 2, \\ |\text{MSS}_4| &= 3, |\text{MSS}_5| = 2, |\text{MSS}_6| = 4, \dots . \end{aligned} \tag{2.6}$$

The last two relations in (2.5) already supplement the basic definition (first relation) in (2.5) by allowing the counting of all MSS roots for relatively small values of n. But the finite set MSS_n to which the MSS roots, ζ_t and ζ_{t+1} corresponding to the leftend of the interval (ζ_t, ζ_{t+1}) belongs is yet to be determined. These MSS roots can also be characterized by the lexical sequence corresponding to a word that gives a closed cycle. Thus, there exists a lexical sequence $\lambda^{(t)}$ such that

$$\zeta(\lambda^{(t)}) = \zeta_t \in \text{MSS}_n, \text{ for some } n \geq 1. \tag{2.7}$$

A rule for obtaining the general lexical sequence $\lambda^{(t)}$ in this result would give a structural result for characterizing the MSS roots, hence, the intervals $(\zeta_t, \zeta_{t+1}]$ without appeal to the computer-generated graphs. Such a result for central sequences is next stated, followed by its proof, followed by some confirming examples:

The central sequence $c_n(t)$ for baseline **B**$_n$ *is given in terms of a supplemental sequence $\Lambda_{n-1}(\lambda^{(t)})$ by the following relation:*

$$c_n(t) = \left(1, \Lambda_{n-1}(\lambda^{(t)})\right), \tag{2.8}$$

where the sequence $\Lambda_{n-1}(\lambda^{(t)})$ is of degree $n-1$.

Proof. Every central sequence must be less than the central sequence $c_n(1) = (1^n)$; hence, must have first part $= 1$, as shown in (2.8). Moreover, it must be the case that

$$\Lambda_{n-1}(\lambda^{(t)}) > (1^{n-1}), \ n \geq 3. \qquad \square \tag{2.9}$$

The rules for obtaining the sequence $\Lambda_{n-1}(\lambda^{(t)})$ begin with the simple division algorithm for the integer $n-1$:

$$\begin{aligned} n-1 &= dm + r, \\ m &= \left\lfloor \frac{n-1}{d} \right\rfloor = \text{greatest integer} \leq \frac{n-1}{d}; d = 1, 2, \dots, n-1, \\ r &\in \{0, 1, \dots, \lfloor \tfrac{n-2}{2} \rfloor\} \text{ (remainder)}, \ n \geq 2. \end{aligned} \tag{2.10}$$

First, a sequence γ is selected from the set of lexical sequences $\mathbb{L}_d$, where this set is any subset of the set $\mathbb{M}_{n-1}$ defined in (1.69):

$$\mathbb{L}_d \subset \mathbb{M}_{n-1} = \{\mathbb{L}_1, \mathbb{L}_2, \dots, \mathbb{L}_{n-1}\}, n \geq 2. \tag{2.11}$$

It is to be noted here that the set $\mathbb{M}_n$ of lexical sequences defined by (1.69) has been replaced by $\mathbb{M}_{n-1}$:

The greatest value of d is $n-1$, and the degree of a lexical sequences $\lambda \in \mathbb{L}_{n-1}$ is $D(\lambda) = n - 2$.

Second, for each $\lambda \in \mathbb{L}_d$ and each $\rho \in \mathbb{A}_r$ a sequence denoted $\Lambda_{n-1}(\lambda; \rho)$ is defined from the elementary division algorithm $n - 1 = md + r$ by the relations:

$$\Lambda_{n-1}(\lambda; \rho) = \begin{cases} (\lambda, 1)^m \, \rho, & \ell(\lambda) \text{ even} \\ (\lambda, -1)^m \, \rho, & \ell(\lambda) \text{ odd}; \end{cases} \tag{2.12}$$

$$\lambda \in \mathbb{L}_d, \rho \in \mathbb{A}_r.$$

The sequences λ and ρ entering this definition are called **divisor** and **remainder** sequences. The notations $(\lambda, 1)$ and $(\lambda, -1)$ in (2.12) denote:

$$\begin{aligned} (\lambda, 1) &= (\lambda_0, \lambda_1, \ldots, \lambda_k, 1), \\ (\lambda, -1) &= (\lambda_0, \lambda_1, \ldots, \lambda_{k-1}, \lambda_k + 1), \\ \ell(\lambda) &= k + 1. \end{aligned} \tag{2.13}$$

These are special examples of a more general rule for the concatenation of two sequences defined in Sect. 3.2 below. The length $\ell(\lambda)$ of a sequence is the number of nonzero parts; the sequence (0) is of length 0, and, if included in a sequence, it does not contribute to its length. *If $\rho = (0)$ in definition (2.12), it is omitted from the sequence.* The degree of the sequence $\Lambda_{n-1}(\lambda; \rho)$ defined in (2.12) is easily checked to be $n - 1$, since $D(\lambda) = d - 1; D(\lambda, \pm 1) = d, D(\rho) = r$.

Each MSS root $\zeta_t \in \mathbb{M}_{n-1}$ must, according to the discussion on MSS roots in Sect. 1.2.4 and on the structure of baseline $\mathbf{B}_n$ in Sect. 1.4 and Sect. 2.1, determine the left endpoint ζ_t of the interval $(\zeta_t, \zeta_{t+1}]$ for which the sequence denoted $c_n(t)$ is central, and this central sequence is unique. Thus, the sets of lexical sequences

$$\{\mathbb{L}_d \,|\, d = 1, 2, \ldots, n - 1\} \tag{2.14}$$

have a principal role. But the lexical sequences that enter into the definition of central sequences $c_n(t)$ in (2.7) for baseline $\mathbf{B}_n$ are yet to be determined. Some preliminary progress can still be made. First, the sequences defined by (2.11)-(2.12) for the unique remainder sequence $\rho \in \mathbb{A}_r$, even though as yet unknown, can be fully ordered, so that

$$\mathbb{C}_n = \{(1 \, \Lambda_{n-1}(\lambda; \rho)) \,|\, \lambda \in \mathbb{M}_{n-1}\}^{ord}. \tag{2.15}$$

The ordering here is to be from greatest-to-least as read left-to-right. It follows then that $\lambda^{(t)}; \rho^{(t)}$ is the t-th sequence in this ordered set

$$\begin{aligned} \mathbb{C}_n &= \{c_n(1), c_n(2), \ldots, c_n(b_n - 1)\} \\ &= \{(1 \, \Lambda_{n-1}(\lambda; \rho) \,|\, \lambda \in \mathbb{M}_{n-1}\}^{ord} \\ &= \{(1 \, \Lambda_{n-1}(\lambda^{(t)}; \rho^{(t)}) \,|\, t = 1, 2, \ldots, b_n - 1\}. \end{aligned} \tag{2.16}$$

The missing ingredient in the above is the identification of the lexical sequences that determine the central sequences. To address this, it is useful to partition the set of sequences $\{\Lambda_{n-1}(\lambda; \rho)\}$ in (2.14)-(2.15), which contains

$|\mathbb{M}_{n-1}| \times |\mathbb{A}_r|$ sequences, into the subsets of divisor sequences $\mathbb{D}_{n-1}(\lambda)$ and remainder sequences $\mathbb{R}_{n-1}(\rho)$ defined as follows:

For each $\lambda \in \mathbb{L}_d$, $d = 1, 2, \ldots, n-1$:

$$\mathbb{D}_{n-1}(\lambda) = \left\{ \Lambda_{n-1}(\lambda; \rho) \,|\, \rho \in \mathbb{A}_{n-1-md} \right\}. \tag{2.17}$$

For each, $\rho \in \mathbb{A}_r$; $r = 0, 1, \ldots, \lfloor \frac{n-2}{2} \rfloor$:

$$\mathbb{R}_{n-1}(\rho) = \left\{ \Lambda_{n-1}(\lambda; \rho) \,\Big|\, \lambda \in \mathbb{L}_d;\; d \in \mathbb{I}_n(r) \right\}; \tag{2.18}$$

$$\mathbb{I}_n(r) = \left\{ d \in 1, 2, \ldots, n-1 \,\Big|\, \lfloor \tfrac{n-1}{d} \rfloor d = n - r - 1 \right\}. \tag{2.19}$$

Thus, the divisor subsets all have the same divisor sequence λ and the remainder sets the same remainder sequence ρ. As these sequences, in turn, assume all possible values as given, respectively, by $\lambda \in \mathbb{L}_d, d = 1, 2, \ldots, n-1$ and $\rho \in \mathbb{A}_r, r = 0, 1, \ldots, \lfloor \frac{n-2}{2} \rfloor$, the full set of all sequences in (2.16) and in (2.17)-(2.18) are exactly enumerated. It is the divisor sets that are of particular interest in resolving the possible multiplicity of ρ-sequences for a selected $\lambda \in \mathbb{L}_d$. Membership in the indexing set $\mathbb{I}_n(r)$ severely limits the lexical sets $\mathbb{L}_d$ admitted into the remainder set $\mathbb{R}_{n-1}(\rho)$:

It is necessary that each division set $\mathbb{D}_n(\lambda)$ contains exactly one remainder sequence ρ of given degree $D(\rho) = r$; otherwise, there can be no unique sequence $\rho(\lambda)$ associated with a selected $\gamma \in \mathbb{L}_d$, $d \in \{1, 2, \ldots, n-1\}$, and the determination of ρ is incomplete.

Certain divisor sets contain but one sequence ρ in their remainder set in consequence of restrictions on r coming from the division algorithm itself. These include:

$$\mathbb{R}_{n-1}(r) = \{ \Lambda_{n-1}(\gamma; r) \,|\, \gamma \in \mathbb{L}_d;\; d \in \mathbb{I}_n(r) \}, \quad r = 0, 1. \tag{2.20}$$

All Λ_{n-1}-sequences for $n = 2, 3, 4, 5$ giving central sequences are included in these two subsets for remainder $r = 0, 1$. It is only for remainders $r \geq 2$ that there are two or more sequences in the divisor set $\mathbb{D}_{n-1}((\rho))$; this occurs for all $D_{n-1}(\rho)$ for all $n \geq 6$. Thus, for $n = 2, 3, 4, 5$, it is always the case that all sequences $\Lambda_{n-1}(\gamma; \rho)$ are uniquely determined by the division algorithm itself. This can be verified explicitly by working-out all five cases, which is left as an exercise.

To deal with remainders $\rho \geq 2$ still new structures must be found. The development of such properties continues.

2.2 Reducible and Irreducible Sequences

A sequence $\Lambda_{n-1}(\lambda; \rho)$ is said to be **reducible** *if it is equal to another sequence of the same form having lesser remainder; that is,*

$$\Lambda_{n-1}(\lambda; \rho) = \Lambda_{n-1}(\lambda'; \rho'),$$

$$D(\rho) = r > r' = D(\rho');\quad \lambda \text{ and } \lambda' \text{ each lexical.} \tag{2.21}$$

A sequence that is not reducible is said to be **irreducible.** Thus, the criterion for central sequences can be phrased in this nomenclature as follows:

Each sequence $\Lambda_{n-1}(\lambda; \rho)$, $\lambda \in \mathbb{L}_d$, $d = 1, 2, \ldots, n-1, \rho \in \mathbb{A}_r$, for which the division algorithm $n - 1 = md + r$ holds, is either irreducible or reducible; if irreducible, it corresponds to a unique central sequence $c_n(\lambda) = (1 \ \Lambda_{n-1}(\lambda; \rho))$, $\mathbb{D}_n(\rho) = r$, with a unique remainder r; otherwise, the sequence is reducible.

The criteria for reducible and irreducible sequences is next developed. First, it is observed that each positive sequence in $\mathbb{A}_r$ can be expressed in one of the following two forms:

$$\text{for each } \beta \in \mathbb{A}_{r-1} = (\beta_0, \beta_2, \ldots, \beta_k), \ D(\beta) = r - 1;$$
$$\rho = (1 \ \beta); \quad \rho = \beta^{+1} = (\beta_0 + 1, \beta_1, \ldots, \beta_k). \tag{2.22}$$

The proof of this relation is by elementary induction on index r: Each sequence is unique, the number is $2 \times 2^{r-2} = 2^{r-1}$, and it holds for $r = 2$.

It now follows that every sequence $\Lambda_{n-1}(\lambda; \rho)$ in (2.21) can be written in one of the following four forms:

$$\ell(\lambda) \text{ even} : \Lambda_{n-1}(\lambda; (1 \ \beta)) \ = \ (\lambda \ 1)^m \ 1 \ \beta; \tag{2.23}$$
$$\ell(\lambda) \text{ even} : \ \Lambda_{n-1}(\lambda; \beta^{+1}) \ = \ (\lambda \ 1)^m \ \beta^{+1}; \tag{2.24}$$
$$\ell(\lambda) \text{ odd} : \Lambda_{n-1}(\lambda; (1 \ \beta)) \ = \ (\lambda \ -1)^m \ 1 \ \beta; \tag{2.25}$$
$$\ell(\lambda) \text{ odd} : \ \Lambda_{n-1}(\lambda; \beta^{+1}) \ = \ (\lambda \ -1)^m \ \beta^{+1}. \tag{2.26}$$

These relations apply to all sequences β of length $\ell(\beta) \geq 2$; they also apply to either parity, even or odd, of $\ell(\beta)$. They are also **complete;** that is, they include every possible sequence of the form $\Lambda_{n-1}(\lambda; \rho), \lambda \in \mathbb{L}_d$, $d = 1, 2, \ldots, n; \rho \in \mathbb{A}_r$, for which the division algorithm $n - 1 = md + r$ holds.

The problem now is to determine which of the forms (2.23)-(2.26) are reducible. The reducibility conditions are next enforced directly from the division algorithm for the $\Lambda_{n-1}(\lambda; \rho)$ sequences and from the forms (2.23)-(2.26) without enforcing yet the lexical condition on λ'. This gives the following **necessary** relations that must hold:

Case $\ell(\lambda)$ even; $\ell(\lambda')$ even: $\hfill (2.27)$

sequence requirement: $(\lambda \ 1)^m \ 1 \ \beta = (\lambda' \ 1)^{m'} \rho', \ D(\rho) > D(\rho'),$

degree requirement: $md + r = m'd' + r'$ and $r > r'$

imply $m'd' > md$ and therefore

$\qquad \lambda' = (\lambda \ 1)^m \alpha$ and $m' = 1$; hence,

$\qquad \alpha \ 1 \ \rho' = (1 \ \beta), \ \alpha = (0)$ or positive.

Case $\ell(\lambda)$ even; $\ell(\lambda')$ odd: $\hfill (2.28)$

sequence requirement: $(\lambda \ 1)^m \beta^{+1} = (\lambda', \ -1)^{m'} \rho', \ D(\rho) > D(\rho'),$

degree requirement: $md + r = m'd' + r'$ and $r > r'$

imply $m'd' > md$ and therefore

$$\lambda' = (\lambda \ 1)^m \alpha' \text{ and } m' = 1; \text{ hence,}$$

$$(\alpha', \ -1) \, \rho' = \beta^{+1}.$$

The conditions for reducibility for the two cases $\ell(\lambda)$ odd; $\ell(\lambda')$ even and $\ell(\lambda)$ odd; $\ell(\lambda')$ odd are obtained from (2.27) and (2.28), respectively, simply by making the replacement of $(\lambda \ 1)^m$ by $(\lambda, \ -1)^m$, with all other relations remaining unchanged, especially, the conditions in the last relation of each of (2.27)-(2.28).

It is still necessary to enforce the rule that the sequence λ' be lexical in (2.27)-(2.28) and in their modification to $\ell(\lambda)$ odd. Sequence lexicality and the associated reverse-lexicographic order rule require:

$\ell(\lambda)$ even:

Conditions that λ' be lexical. For each $k = 0, 1, \ldots, m - 1$: (2.29)

$(\lambda \ 1)^{m-k} \alpha > \alpha$, all k even; $(\lambda \ 1)^{m-k} \alpha < \alpha$, all k odd.

$\ell(\lambda)$ odd:

Conditions that λ' be lexical. For each $k = 0, 1, \ldots, m - 1$: (2.30)

$(\lambda, \ -1)^{m-k} \alpha > \alpha$, all k even; $(\lambda, \ -1)^{m-k} \alpha < \alpha$, all k odd.

These conditions for lexicality must hold for all k even and all k odd that are in the domain $k \in \{0, 1, \ldots, m-1\}$. It is also allowed to take $\alpha = (0)$ in each relation (2.29)-(2.30). The lexical conditions then require $(\lambda, \ 1)^{m-k} > (0)$ for k even **and** $(\lambda, \ 1)^{m-k} < (0)$ for k odd. This is a contradiction unless $m = 1$, in which case the second condition is inapplicable since only $k = 0$ is allowed. The conclusions are:

> If λ is lexical and $\ell(\lambda)$ is even, necessary and sufficient
> conditions that the sequence $(\lambda \ 1)^m \ \alpha$ be lexical are
> $m = 1$ and that $\lambda \ 1 \alpha$ be lexical; (2.31)
>
> if λ is lexical and $\ell(\lambda)$ is odd, necessary and sufficient
> conditions that the sequence $(\lambda, \ -1)^m \alpha$ be lexical are
> that $m = 1$ and that $(\lambda \ - 1) \ \alpha$ be lexical. (2.32)

The result (2.31) holds even for $\alpha = (0)$, in which case it is always true that $(\lambda \ 1), \ell(\lambda)$ even is always lexical, even for $\lambda = (0)$. But the condition λ lexical for $\ell(\lambda)$ odd in (2.32) is not sufficient for $(\lambda, \ -1) \ \alpha$ to be lexical, even for $\alpha = (0)$; hence, it must be specified in addition to $m = 1$, as shown.

The condition of lexicality of λ' in the reduction formula (2.31) severely restricts the λ that meet the reduction criterion: It leaves as irreducible all sequences $\Lambda_{n-1}(\lambda; \rho)$ for which the division algorithm $n - 1 = md + r$ admits sequences with $m \geq 2$ and $r \geq 2$. This first occurs for $n = 9$, in which case $m = 2, d = 3, r = 2$. These parameters give the following two cases, each of which is irreducible:

$$\Lambda_8((2); (2)) = (3\ 3\ 2); \quad \Lambda_8((2); (1\ 1)) = (3\ 3\ 1\ 1). \tag{2.33}$$

Thus, neither of these sequences can be written in the form $\Lambda_8(\lambda'; \rho')$ with λ' lexical and $\rho' = (0)$ or (1). The uniqueness of every sequence $\Lambda_{n-1}(\lambda; \rho)$ required by the condition that the sequence $\lambda \in \mathbb{L}_d$ is the lexical sequence such that the MSS root $\zeta(\lambda)$ has the property that $\lfloor \zeta(\lambda) \rfloor < \zeta(\lambda)$ is not resolved by the reducible criterion alone for every allowed $\Lambda_{n-1}(\lambda; \rho)$. But now the fact that each sequence in the set $\{(1\ \Lambda_{n-1}(\lambda; \rho) \,|\, \rho \in \mathbb{A}_r\}$ is created synchronously with the central sequence comes into play: The least of these sequences must be the central sequence; it is the sequence $\rho = (r)$ for m even, and the sequence $\rho = (1\ r - 1)$ for m odd. Thus, the following result holds:

$$
\begin{aligned}
m \text{ even} \;:\; & c_n(\lambda) = (1\ \Lambda_{n-1}(\lambda; (r))) \\
& < (1\ \Lambda_{n-1}(\lambda; \rho), D(\rho) = r, \rho < (r); \\
m \text{ odd} \;:\; & c_n(\lambda) = (1\ \Lambda_{n-1}(\lambda; (1\ r - 1))) \\
& < (1\ \Lambda_{n-1}(\lambda; \rho), D(\rho) = r, \rho > (1\ r - 1).
\end{aligned}
\tag{2.34}
$$

Thus, not only are the sequences $c_n(\lambda; (r))$, all m even, and the sequences $c_n(\lambda; (1\ r-1))$, all m odd, the central sequences created at the MSS root $\zeta(\lambda)$ for $\ell(\lambda)$ even, but also the remaining sequences in the sets (2.28) and (2.29), respectively, are created syncronously with the central sequence; moreover, they are the adjacent sequences just above the central sequence in the baseline $\mathbf{B}_n$. They are adjacent in $\mathbb{A}_n$ because it is easily shown that no sequence in $\mathbb{A}_n$ can fall between any pair.

Examples: Two examples that validate the above classification of sequences of the form $\Lambda_{n-1}(\lambda; \rho)$ are:

1. $n = 10,\ d = 3,\ m = 2,\ r = 3$:

$$
\begin{aligned}
\Lambda_9((2); (3)) &= (3\ 3\ 3); \\
\Lambda_9((2); (2\ 1)) &= (3\ 3\ 2\ 1); \\
\Lambda_9((2); (1^3)) &= (3\ 3\ 1\ 1\ 1); \\
\Lambda_9((2); (1\ 2)) &= (3\ 3\ 1\ 2).
\end{aligned}
\tag{2.35}
$$

2. $n = 12,\ d = 4,\ m = 2,\ r = 3$:

$$
\begin{aligned}
\Lambda_{11}((2\ 1); (3)) &= (2\ 1\ 1\ 2\ 1\ 1\ 3); \\
\Lambda_{11}((2\ 1); (2\ 1)) &= (2\ 1\ 1\ 2\ 1\ 1\ 2\ 1); \\
\Lambda_{11}((2\ 1); (1^3)) &= (2\ 1\ 1\ 2\ 1\ 1\ 1\ 1\ 1); \\
\Lambda_{11}((2\ 1); (1\ 2)) &= (2\ 1\ 1\ 2\ 1\ 1\ 1\ 2).
\end{aligned}
\tag{2.36}
$$

Each of these sequences has $m = 2$ and is irreducible, as directly verified: There exists no lexical sequence λ' and remainder $r' < 3$ for which any of these sequences is reducible. Since $m = 2$, the central sequence is the one with greatest remainder; namely, $\rho = (3)$, and the remaining three occur immediately above the central sequence, which is $c_{10}((2);(3)) = (1 \; \Lambda_9((2);(3)))$ in Example 1, and $c_{12}((2\;1);(3)) = (1 \; \Lambda_{11}((2\;1);(3))$ in Example 2, all in proper order and adjacent, and in the same column of baseline $\mathbf{B}_{10}$ and baseline $\mathbf{B}_{12}$, respectively, since they are all created at the respective MSS roots $\zeta((2))$ and $\zeta((2\;1))$.

Regrettably, in order to reduce the number of inverse graphs given in Chapter 7 to a more manageable number, the computer-generated graphs for $P\,10$ and $P\,12$ were not included, although they were computer-calculated by the same methods used for all such graphs, and (2.35)-(2.36) were verified to be correct. Inverse graphs for $P\,16$ are included in Chapter 7 as representative of the intricate ζ-evolution of inverse graphs for higher n.

The MSS roots for $n = 2 - 8$ at which the successive central sequences are created are obtained from the above results on irreducible sequences:

$$\mathbb{C}_2 = \{\zeta((0))\}$$
$$\mathbb{C}_3 = \{\zeta((0)), \zeta((1))\},$$
$$\mathbb{C}_4 = \{\zeta((0)), \zeta((1)), \zeta((2))\},$$
$$\mathbb{C}_5 = \{\zeta((0)), \zeta((1)), \zeta((2\;1)), \zeta((2)), \zeta((3))\},$$
$$\mathbb{C}_6 = \{\zeta((0)), \zeta((1)), \zeta((2\;1)), \zeta((2\;1\;1)), \zeta((2)),$$
$$\zeta((3\;1)), \zeta((3)), \zeta((4))\}, \tag{2.37}$$
$$\mathbb{C}_7 = \{\zeta((0)), \zeta((1)), \zeta((2\;1)), \zeta((2\;1^3)),$$
$$\zeta((2\;1^2)), \zeta((2)), \zeta((3\;2)), \zeta((3\;1)), \zeta((3\;1\;1)), \zeta((3)),$$
$$\zeta((4\;1)), \zeta((4)), \zeta((5))\},$$
$$\mathbb{C}_8 = \{\zeta((0)), \zeta((1)), \zeta((2\;1)), \zeta((2\;1^3)),$$
$$\zeta((2\;1^5)), \zeta((2\;1^4)), \zeta((2\;1^2)), \zeta((2\;1\;2\;1)), \zeta((2)),$$
$$\zeta((3\;2)), ((3\;2\;1)), \zeta((3\;1)), \zeta((3\;1^2)), \zeta((3\;1^3)), \zeta((3\;1\;2)),$$
$$\zeta((3)), \zeta((4\;1)), \zeta((4\;1^2)), \zeta((4)), \zeta((5\;1)), \zeta((5)), \zeta((6))\}.$$

The number of sequences in these respective sets is:

$$|\mathbb{C}_2| = 1, |\mathbb{C}_3| = 2, |\mathbb{C}_4| = 3, |\mathbb{C}_5| = 5, |\mathbb{C}_6| = 8, |\mathbb{C}_7| = 13, |\mathbb{C}_8| = 22. \tag{2.38}$$

These creation-value MSS roots do not include the primordial sequence (n), which is never an MSS root. The central curve $C_\zeta^n\left((n)\,\middle|\,\overline{(n)}\right)$ is created at the left endpoint $b_0 = 0$ of baseline $\mathbf{B}_n$.

It will be observed that for general n exactly one central sequence is created at the MSS root given by each lexical sequence in the set $\mathbb{M}_{n-1}$. This implies that the number of columns in baseline $\mathbf{B}_n$ is given by

$$b_n = 1 + \sum_{d=1}^{n-1} |\mathbb{L}_d|; \tag{2.39}$$

$$b_n = b_{n-1} + |\mathbb{L}_{n-1}|, \; b_1 = 1, \; n \geq 2. \qquad (2.40)$$

Thus, from relation (1.70) for the number of lexical sequences, the number of columns b_n in baseline $\mathbf{B}_n$ is known in closed form in many cases and recursively for the remaining.

The following result can now be asserted. It is placed in a box for prominence of display:

> There exists only one creation table $\mathbb{T}_n$ with the property that the sequences that appear in the same column as the central sequence are unique. $\qquad (2.41)$

What is incomplete in the assertion (2.41) about the creation table $\mathbb{T}_n$ is the rules that determine the unique sequences that constitute the sequences that appear in the same column as a given central sequence: their uniqueness is already assured. The validity of this result can be checked explicitly from the computer-generated graphs given in Chapter 7. But it is desirable to know exactly which sequences from $\mathbb{A}_n$ go into each column with a given central sequence. Since this result is one of the more important ones given in this monograph, it is useful to have another perspective of its structure, which at the same time serves as a proof.

The method of addressing the above issue is based directly on the properties of branch functions. Let $\alpha = (\alpha_0, \alpha_1, \ldots, \alpha_k)$ and $\beta = (\beta_0, \beta_1, \ldots, \beta_j)$ denote arbitrary positive sequences of lengths $k + 1$ and $j + 1$. Then, the branch functions satisfy the following function composition rules for the concatenation of sequences based on the correspondence with two letters (see relations (1.47)-(1.53)):

$$\Psi_\zeta(\alpha\beta; x) \;=\; \Psi_\zeta\Big(\alpha; \Psi_\zeta(\beta; x)\Big); \qquad (2.42)$$

$$\Psi_\zeta(\alpha\overline{\beta}; x) \;=\; \Psi_\zeta\Big(\alpha; \Psi_\zeta(\overline{\beta}; x)\Big); \qquad (2.43)$$

$$\alpha\beta \;=\; (\alpha_0, \alpha_1, \ldots, \alpha_k, \beta_0, \beta_1, \ldots, \beta_j); \qquad (2.44)$$

$$\alpha\overline{\beta} \;=\; (\alpha_0, \alpha_1, \ldots, \alpha_k + \beta_0, \beta_1, \ldots, \beta_j). \qquad (2.45)$$

The application of relations (2.42)-(2.45) made here is to the two distinct classes of functions for which $\alpha = (1)$, and for which

$$\alpha > \alpha' \text{ implies } \alpha^{+1} > \alpha'^{+1}; \text{ and } (1 \; \alpha) < (1 \; \alpha'). \qquad (2.46)$$

Application of the concatenation formulas (2.42)-(2.45) now gives the following important relations between branch functions:

$$\Psi_\zeta(\alpha^{+1}; x) \;=\; \Psi_\zeta\Big((1); \Psi_\zeta(\overline{\alpha}; x)\Big), \; \alpha \in \mathbb{T}_{n-1}; \qquad (2.47)$$

$$\Psi_\zeta((1\ \alpha); x) \;=\; \Psi_\zeta\Big((1); \Psi_\zeta(\alpha; x)\Big),\ \alpha \in \mathbb{T}_{n-1}. \qquad (2.48)$$

The sequences that appear in the Ψ-functions on the left are all in $\mathbb{T}_n$, while on the right only the simplest of the inverse function occurs, namely,

$$\Psi_\zeta((1); x) = 1 + \sqrt{1 - \frac{x}{\zeta}}, \qquad (2.49)$$

which is evaluated at the x-values given, respectively, by $\Psi_\zeta(\overline{\alpha}; x)$ and $\Psi_\zeta(\alpha; x)$. Relations (2.47)-(2.49) now imply relations (2.50)-(2.53) below:

1. The branch function $\Psi_\zeta(\alpha^{+1}; x)$, $\alpha^{+1} \in \mathbb{T}_n$ is real, if and only if the conjugate branch function $\Psi_\zeta(\overline{\alpha}; x)$, $\alpha \in \mathbb{T}_{n-1}$ is real, and the following relation is satisfied (the condition $(\zeta, x) \in \mathbb{R}^2$ are always implicit, unless otherwise specified):

$$\Psi_\zeta(\overline{\alpha}; x) \leq \zeta. \qquad (2.50)$$

For all $(\zeta, x) \in \mathbb{R}^2$ that satisfy this relation, the real branch function $\Psi_\zeta(\alpha^{+1}; x)$ satisfies:

$$1 \leq \Psi_\zeta(\alpha^{+1}; x) \leq 2. \qquad (2.51)$$

2. The branch function $\Psi_\zeta((1, \alpha); x)$, $(1, \alpha) \in \mathbb{T}_n$ is real, if and only if the branch function $\Psi_\zeta(\alpha; x)$, $\alpha \in \mathbb{T}_{n-1}$, is real, and the following two relations are satisfied:

$$\Psi_\zeta(\alpha; x) \leq \zeta \text{ and } x \leq \zeta. \qquad (2.52)$$

For all $(\zeta, x) \in \mathbb{R}^2$ that satisfy this relation, then the real branch function $\Psi_\zeta((1, \alpha); x)$ satisfies:

$$1 \leq \Psi_\alpha((1, \alpha); x) \leq 2. \qquad (2.53)$$

Conditions (2.47)-(2.53) are exact in specifying exactly the properties that the branch functions $\Psi_\zeta(\alpha; x)$, $\alpha \in \mathbb{A}_{n-1}$, must possess in order to yield the domains for which the branch functions in the inverse graph G_ζ^n are real. Of course, the placement of α in Table $\mathbb{T}_{n-1}$ already gives the domain for which $\Psi_\zeta(\alpha; x)$ is real, but conditions (2.47)-(2.53) go beyond this. Indeed, these conditions must yield the characteristic columns of baseline $\mathbf{B}_n$ from those in baseline $\mathbf{B}_{n-1}$; that is, relations (2.47)-(2.53) contain implicitly the placement of each sequence in $\mathbb{T}_{n-1}$ into its characteristic column in $\mathbf{B}_n$.

Summary: The $+1$-rule and the $(1\ \alpha)$-rule given in relations (2.47)-(2.48) are basic to the construction of Table $\mathbb{T}_n$ from Table $\mathbb{T}_{n-1}$; the application of each of these rules to the 2^{n-2} positive branch functions in $\mathbb{T}_{n-1}$ gives all the 2^{n-1} positive branch functions in $\mathbb{T}_n$. But it is the composition relations (2.47)-(2.49) **and the resulting conditions (2.50)-(2.53)** on Ψ-functions for sequences in $\mathbb{A}_{n-1}$ that provide the needed information to obtain the column mapping between the characteristic columns of baseline $\mathbf{B}_{n-1}$ to the characteristic columns of baseline $\mathbf{B}_n$:

Each label in a given characteristic column of baseline $\mathbf{B}_{n-1}$ of Table $\mathbb{T}_{n-1}$ is assigned to a unique characteristic column of baseline $\mathbf{B}_n$ of Table $\mathbb{T}_n$ in consequence of the enforcement of the reality conditions stated in relations

(2.51) and (2.53) on the branch function relations (2.47)-(2.48). **But this is exactly the set** $\mathbb{C}_n^*$, as next shown.

The mapping rule for columns has the following form:

$$\alpha \in Col_{t'}^{(n-1)} \quad \mapsto \quad \begin{cases} \alpha^{+1} \in Col_{t_1}^{(n)}, \\ (1\ \alpha) \in Col_{t_2}^{(n)}; \end{cases} \tag{2.54}$$

$$t' \;\; = \;\; 0, 1, \ldots, b_{n-1}; \; t_1, t_2 \in \{0, 1, \ldots, b_n\}.$$

Condition $\Psi(\overline{\alpha}; x) \leq \zeta$ in (2.50) always implies that $1 \leq \Psi_\zeta(\alpha^{+1}; x) \leq 2$ in (2.51) is always satisfied for $\Psi(\overline{\alpha}; x)$ real, since the latter always falls between 0 and 1, when real. The situation is more complicated for relation (2.52): It is possible to have $\Psi_\zeta(\alpha; x)$ real and $\Psi_\zeta(\alpha; x) \geq \zeta$, in which case $\Psi_\zeta((1\ \alpha); x)$ is still complex; hence, it is necessary to still enforce the condition $\Psi_\zeta(\alpha; x) \leq \zeta$. But this is exactly the reason the set $\mathbb{C}_n^*$ was introduced. The structures described above lead exactly to $\mathbb{C}_n^*$, the set of deterministic-computer-generated inverse graphs.

The results given above are now given the same box prominence as the existence result (2.41):

> The positive sequences that appear in the same column in the baseline $\mathbf{B}_n$ constitute exactly the set of irreducible sequences, which, in turn, is exactly the set $\mathbb{C}_n^*$ of deterministic-computer-generated inverse graphs.   (2.55)

Chapter 3

Description of Events in the Inverse Graph

It is now quite reasonable to regard the collection of **deterministic chaos events** as described in this monograph as a **Complex Adaptive System,** where for our purposes we defined a Complex Adaptive System to be one in which there are only a few undefined "objects" from which follows a set of properties of the objects that are rich in structure, and even often, unexpected. In this Chapter, this definition of a Complex Adaptive System is supplemented still further with examples of the richness of structure of the system. Many of these detailed features of the collection of inverse graphs can be observed qualitatively from the computer-generated inverse graphs themselves. These include the two kinds of bifurcation events, the so-called tangent bifurcations and period-doubling bifurcations that precede the value of the parameter ζ at which new branches are created. It also includes the description of their dynamical shape and p-curves that constitute the full graph at each value of the parameter ζ; including the left and right motions of each p-curve. The shape evolution of the inverse graph in the two parameters ζ and x of the underlying parabola is quite vivid in the deterministic-computer-generated inverse graphs $\mathbb{C}_n^*$ given in Chapter 7, but the events taking place in these inverse graphs are still nontrivial to identify. Progress toward the stated goal of proving properties of $\mathbb{C}_n^*$ is less than obvious.

3.1   Domains of Definition of Branches and Curves

The underlying problem can be quite easily visualized and stated:

The left and right extremal coordinates $x_\zeta^{(1)}(\alpha; x) \leq x_\zeta^{(2)}(\alpha; x)$ of each branch of $\Psi_\zeta(\alpha; x)$ that appears in the inverse graph at a given value of ζ are the left-most and right-most points of the branch. The problem is to determine these extremal coordinates.

The process of determining the extremal coordinates is initiated by introducing the following set of sequences:

$$\mathcal{A}^{n-1} = \left(\bigcup_{m=1}^{n-1} \mathbb{A}_m \right)^{ord}, \; n \geq 2, \qquad (3.1)$$

where, by definition, $\mathbb{A}_1 = \{(n-1), (n-2), \ldots, (1)\}$, and $\mathbb{A}_m, m \geq 2$, is the set of all positive α sequences that add to m. The sequences in $\mathcal{A}^{n-1}$ are to be ordered from the greatest sequence $(n-1)$ to the least sequence (1). The set $\mathcal{A}^{n-1}$ is **complete** in the sense that every sequence of order 1 to order $n-1$ is contained in $\mathcal{A}^{n-1}$. Thus, $\mathcal{A}^{n-1}$ contains, in all, $2^{n-1} - 1, n \geq 2$, distinct positive sequences. This means that at $\zeta = 2$, where all 2^{n-1} positive graphs $\alpha \in \mathbb{A}_n$ of degree n are present in the set of inverse graphs, the $2^{n-1} - 1$ labels in the set $\mathcal{A}^{n-1}$ can be chosen to enumerate the boundary lines between the sequences in the set $\mathbb{A}_n$, where the missing sequence (0) is assigned to the central $y = 1$ line. This exact matching of sequences makes it natural to use these sequences not only to label the branch functions in the inverse graph, but also to label the boundary lines defined by the extremal points of each inverse graph, thus obtaining a quite vivid picture of each inverse graph as a distinct and unique object. This is fully illustrated in the collection of inverse graphs $P\,4$ in Chapter 7 at the value $\zeta = 2.00000$. Moreover, the boundary lines at any selected value of ζ are obtained simply by deleting the boundary labels for those branches of the inverse graph not yet created. The proof of the general result is simply one of counting the number of boundaries that are needed to account for the number of inverse graphs present at the selected value of ζ. It follows that:

> For each pair $\alpha, \alpha' \in \mathbb{A}_n$ with $\alpha > \alpha'$, there exists a unique
> sequence $\beta \in \mathcal{A}^{n-1}$ such that $\alpha > \beta > \alpha'$; the sequence β is   (3.2)
> the unique boundary sequence at ζ.

It is very important to keep in mind that by their very definition extremal values are unique. This means: Any rule that assigns distinct sequences to the extremal values, one sequence to each extremal value can be used to enumerate them. This is what has been done just above (3.2), using the natural choice of α sequences upon which this monograph is based. That this should occur is somewhat of a surprise! (As often the case, statements are made about positive sequence, but it is always intended that they extend to conjugate sequences as well by the standard rule.)

Despite the all inclusiveness of the above choice, it is still useful to see how the method works in an example:

Example. $n = 4$: At $\zeta = 2$, the following order relation between sequences holds:

$$\begin{aligned}
(4) \;\; &> \;\; (3) > (3\ 1) > (2) > (2\ 1\ 1) > (2\ 1) > (2\ 2) \\
&> \;\; (1) > (1\ 1\ 2) > (1^3) > (1^4) > (1\ 1) \\
&> \;\; (1\ 1\ 2) > (1\ 2) > (1\ 3) > (0).
\end{aligned} \qquad (3.3)$$

For $\zeta \in (0, \zeta(0)]$, the following order relation between sequences holds:

$$(4) > (0). \tag{3.4}$$

For $\zeta \in (\zeta(0), \zeta(1)]$, the following order relation between sequences holds:

$$(4) > (3) > (3\ 1) > (2) > (2\ 1\ 1) > (1) > (1^4) > (0). \tag{3.5}$$

For $\zeta \in (\zeta(2), \zeta(3)]$, the following order relation between sequences holds:

$$
\begin{aligned}
(4) \ &> \ (3) > (3\ 1) > (2) > (2\ 2\ 1) > (2\ 1) > (2\ 2) > (1) \\
&> \ (1\ 1\ 2) > (1^3) > (1^4) > (1\ 1) > (1\ 2\ 1) > (0).
\end{aligned} \tag{3.6}
$$

The sequences $\alpha \in \mathbb{A}_4$ are known for each ζ in the four baseline intervals of $\mathbf{B}_4$:

$$
\begin{aligned}
(0, \zeta(1)] : \quad & \text{the sequence is } (4); \\
(\zeta(1), \zeta(2)] : \quad & \text{the sequences are } (4) > (3\ 1) > (2\ 1\ 1) > (1^4); \\
(\zeta(1), \zeta(2)] : \quad & \text{the sequences are } (4) > (3\ 1) > (2\ 1\ 1) > (2\ 2) \\
& > (1\ 1\ 2) > (1^4) > (1\ 2\ 1); \\
(\zeta(2), \infty) : \quad & \text{the sequences are } (4) > (3\ 1) > (2\ 1\ 1) > (2\ 2) \\
& > (1\ 1\ 2) > (1^4) > (1\ 2\ 1) > (1\ 3).
\end{aligned} \tag{3.7}
$$

Since the α sequences given by (3.7) are exactly the sequences that appear in the positive inverse graph for the indicated baseline interval in $\mathbf{B}_4$, the remaining sequences in each of (3.4)-(3.6) are the boundary sequences given by: (0) for interval $(0, \zeta(1)]$; (3) > (2) > (1) > (0) for interval $(\zeta(1), \zeta(2)]$ for interval $(\zeta(2), \infty)$. Thus, rather elementary rules determine uniquely all boundary sequences from the fully determined α-sequences present in the inverse graph at each value of ζ. It has now been shown that the full system consisting of the creation values of all branches in the inverse graph, and their assigned boundary sequences, is known explicitly:

The deterministic computer-based **Chaos Theory** *developed in this monograph based on properties of the inverse graph and the combinatorics of words on two letters is a* **Complex Adaptive System.**

3.2 Concatenation, Harmonics, and Antiharmonics

The product or concatenation $\alpha\beta$ and $\alpha\bar{\beta}$ of two arbitrary positive sequences of length $k+1$ and $m+1$ given by $\alpha = (\alpha_0, \alpha_1, \ldots, \alpha_k)$ and $\beta = (\beta_0, \beta_1, \ldots, \beta_m)$ is defined by

$$
\alpha\beta \ = \ (\alpha_0, \alpha_1, \ldots, \alpha_k, \beta_0, \beta_1, \ldots, \beta_m),
$$
$$\tag{3.8}$$
$$
\alpha\bar{\beta} \ = \ (\alpha_0, \alpha_1, \ldots, \alpha_{k-1}, \alpha_k + \beta_0, \beta_1, \ldots, \beta_m).
$$

The corresponding inverse functions satisfy the relations:

$$\Psi_\zeta(\alpha\beta; x) = \Psi_\zeta(\alpha; \Psi_\zeta(\beta; x)), \quad \Psi_\zeta(\alpha\bar{\beta}; x) = \Psi_\zeta(\alpha; \Psi_\zeta(\bar{\beta}; x)). \tag{3.9}$$

The harmonic sequence $h(\alpha)$ and antiharmonic sequence $a(\alpha)$ associated with a given positive sequence $\alpha = (\alpha_0, \alpha_1, \ldots, \alpha_k)$ of length $k+1$ are defined, respectively, by

$$h(\alpha) = \begin{cases} (\alpha, 1)\alpha, & k \text{ odd}, \\ (\alpha, -1), & k \text{ even} \end{cases}, \quad a(\alpha) = \begin{cases} (\alpha, -1)\alpha, & k \text{ odd}, \\ (\alpha, 1)\alpha, & k \text{ even} \end{cases}, \quad (3.10)$$

where $(\alpha, -1) = (\alpha_0, \alpha_1, \ldots, \alpha_{k-1}, \alpha_k + 1)$.

3.3 Fixed Points as Dynamical Objects

Fixed points were illustrated in Sect. 1.2.2 as dynamical objects, mostly for the moving point $2 - \frac{1}{\zeta}$. But, of course, there are many fixed points present in the inverse graph G_ζ^n and the graph H_ζ^n as ζ runs over all values in the baseline interval $[0, 2]$. A suitable definition for all inverse graphs for arbitrary n is introduced in this section.

Let the parameter $\zeta \in (0, 2)$ be specified, but arbitrary. A set of points $\{x_1, x_2, \ldots, x_r\}$ obtained by the sequential iteration $x_{i+1} = p_\zeta(x_i) = \zeta x_i(2 - x_i), i = 1, 2, \ldots, r$, of an initially chosen point x_1 such that the condition $x_{r+1} = x_1$ holds is called an r-cycle of the parabola $p_\zeta(x)$. The point x_1 thus satisfies $p_\zeta^r(x_1) = x_1$; it is called a **fixed point** of the r-th iteration of the parabola $p_\zeta(x)$. But then it is also the case that $p_\zeta^r(x_i) = x_i$, for each $i = 1, 2, \ldots, r$; hence, each point x_i in the original set with x_1 is a fixed point of the parabola $p_\zeta(x)$. Each such point is also a real root of the polynomial equation:

$$p_\zeta^r(x) - x = 0. \quad (3.11)$$

This is a polynomial relation of degree 2^r with coefficients that are themselves polynomials in the parameter ζ. The same r-cycle is effected by each of the iterations $x_{j+1} = p_\zeta(x_j)$, where j is any member of the cyclic set of values $j = k, k+1, \ldots, r, 1, 2, \ldots, k-1$, each $k = 2, 3, \ldots, r$. Thus, if the correspondence with a single word to any member of an r-cycle is known, then the word for each member of the r-cycle is obtained by the cyclic permutations of the letters constituting the word. There is no reference whatsoever in this definition of an r-cycle to the inverse graph: It is purely a combinatorial property of words on two letters.

The relation of the points of the inverse graph G_ζ^n to fixed points as defined above comes about because fixed points are the **invariant points** (the same (x, x)-coordinates) between the graph H_ζ^n and its inverse G_ζ^n. (Fixed points for the parabolic map could have been defined in this manner.) But it is also the case that fixed points are created whenever a p-curve meets (becomes tangent to) the $45°$-line.

These facets of sets of fixed points just described, their cyclic permutation property, their invariant property, and their creation property must all be woven together in a consistent manner to give a full picture of the dynamics of fixed points. There is little hint in the separate properties as to how this is to be achieved.

The direct determination of r-cycles is quite a difficult procedure because fixed points are dynamical objects; that is, each fixed point is a smooth function $x_i = x_i(\zeta), i = 1, 2, \ldots, r$ of the parameter ζ, a function that carries the fixed point smoothly along the $45°$-line and always belongs to a branch of the inverse graph G_ζ^n. The properties of the inverse graph bringing the dynamics of fixed points into existence is not in evidence.

A closer look at the deterministic-computer-generated inverse graphs $\mathbb{C}_n^*$ in Chapter 7 reveals that all fixed points originate from a branch of a p-curve becoming tangent to the $45°$-line. This tangency property is a key for explaining the origin of dynamical fixed points. Indeed, the inverse graphs $\mathbb{C}_n^*$ in Chapter 7 show exactly this smooth motion of all fixed points, including the important case for $n = 1$, which has $p_\zeta(x) = x$; that is, $x = 2 - \frac{1}{\zeta}$ for all $x > 0$. This is just the fixed point already discussed in Sect. 1.2.2.

But it is also observed that following the tangency event there always occurs in $\mathbb{C}_n^*$ still another important event, called a **bifurcation event** in which two new fixed points are created. It is in these bifurcation events that the origin of fixed points is to be found.

3.4   The Fabric of Bifurcation Events

A bifurcation event is said to have occurred in the inverse graph whenever there is a change in the number of fixed points. Since fixed points are invariants between the inverse graph and the original graph H_ζ^n itself, it is important to understand the ζ-evolution of bifurcation events; that is, how a bifurcation event manifests itself in the inverse graphs $\mathbb{C}_n^*$ presented in Chapter 7.

It is a well-known result that for the parabolic map there are two types of bifurcation events, saddle-node and period-doubling. *Each creates two new fixed points.*

A saddle-node bifurcation is one that creates two new fixed points by the motion of a p-curve approaching the $45°$-line from the left or the right, becoming exactly tangent, and then simply moving across the $45°$-line. These are the only bifurcations that can occur in the inverse graph for n odd.

A period-doubling bifurcation is one that creates two new fixed points by a rather intricate motion of an existing p-curve about an existing fixed point already on the p-curve. The motion may be described as a "propeller-like" motion around the existing fixed point. Both period-doubling and saddle-node bifurcations occur for even n in a highly organized way, yet to be described.

Pictures of a saddle-node bifurcation are presented for $n = 3$ in Chapter 7 on the three inverse graphs $P3$ for $\zeta = 1.91200, 1.91400, 2.00000$. The left-moving central p-curve $C_\zeta^2((1\ 2)\,|\,\overline{(1\ 2)})$ simply moves across the $45°$-line creating a single fixed point at exactly the point of tangency, this point belonging to the upper branch labeled $((1\ 2))$ of the central curve. This fixed point immediately splits into two fixed points at the tangency point, and these two fixed points move smoothly apart, with the upper point moving onto the upper branch (1 2) of the p-curve, the lower point onto the lower conjugate branch $\overline{(1\ 2)}$. These two fixed points remain on these respective **residency branches** (1 2) and $\overline{(1\ 2)}$ for all greater ζ. Obvious modifications are to be made for a right-moving saddle-node bifurcation. This is the standard picture for a saddle-node bifurcation event, even when it does not occur on a central p-curve, where now the labels of the upper and lower branches of the p-curve are a label τ and its complement $\tilde{\tau}$, which are adjacent sequences in G_ζ^n. There is one **exception,** which is the primordial event, as described above, where the saddle-node bifurcation was accompanied by an unavoidable primordial fixed point on the conjugate graph.

Saddle-node bifurcations are easily recognized; for odd n; they must occur in succession, one in each successive baseline interval, until all labels are fully assigned in the creation table $\mathbb{T}_n$. The anatomy of this situation for even n is much more intricate: *It remains to show how saddle-node bifurcations and period-doubling bifurcations fall in place for even n.*

3.5   The Anatomy of Period-Doubling Bifurcations

A period-doubling bifurcation event always takes place by the motion of a p-curve around an existing fixed point, accompanied by the creation of two new fixed points emerging out of the original.

It is instructive to describe the full ζ-evolution of bifurcations for $n = 4$ in the inverse graph G_ζ^4 as ζ increases from 0 to ∞. This serves as a prototype for all even n. But, first, it is useful to describe the period-doubling bifurcation event for the primordial interval $(0, 1]$, since the details of its evolution can be described, once and for all, for all even n. This general result for n for the interval $(0, 1]$ is next given, followed by the special results for the full interval $(\zeta, 2]$ for $n = 4$:

Interval $(0, 1]$. Arbitrary n: The dynamical moving fixed point $x(\zeta) = 2 - \frac{1}{\zeta}$ emerges from the origin $(0, 0)$ and proceeds along the 45°-line on the lower branch of central graph $\mathcal{C}_\zeta^n \left((n) \, \middle| \, \overline{(n)} \right)$ until it meets the central point $(1, 1)$ of the graph; here it moves onto the upper branch (1^n) of the central graph $\mathcal{C}_\zeta^n \left((1^n) \, \middle| \, \overline{(1^n)} \right)$, which is its **permanent branch of residency,** even after this central graph splits apart when ζ meets the creation point for the next central graph.

The description of fixed-point creation for the case $n = 4$ continues.

3.5.1   The Complete Description for $n = 4$

Interval $(0, 2]$. $n = 4$: For the interval $(0, 1]$, the motion of the dynamical fixed point $x(\zeta) = 2 - \frac{1}{\zeta}$ is that described by setting $n = 4$ in the result stated above for general n, as follows:

Interval $(0, 1]$: For $\zeta \in (0, 1]$, the primordial p-curve denoted by

$$C_\zeta^4 \left((4) \,\middle|\, \overline{(4)} \right) \tag{3.12}$$

is the only curve present in the inverse graph. Its motion through the interval generates the following fixed-point events: For $\zeta \in (0, 1/2]$, the stationary fixed point $(0, 0)$ is the only one present. At $\widehat{\zeta} = 1/2$, the primordial p-curve (3.12) becomes exactly tangent to the 45°-line and, as ζ increases, it moves smoothly through the central point $(1, 1)$ onto the upper branch of the central p-curve $C_\zeta^4 \left((1^4) \,\middle|\, \overline{(1^4)} \right)$, which is its permanent branch of residency, even after this central graph splits apart at the creation point of the next central interval $C_\zeta^4 \left((1\ 2\ 1) \,\middle|\, \overline{1\ 2\ 1} \right)$ at $\zeta((1))$ (see (5.32), which also shows the directions of motion of the central p-curves).

Interval $(1, \zeta((1))]$: A period-doubling bifurcation is initiated at the tangency point $\widehat{\zeta} = 3/2$. It is described as follows: The motion is around the tangent point that resulted as the motion of the moving fixed point $\overline{(4)}$ crossed through the central point $(1, 1)$ of the graph onto the central branch (1^4), where it inherits the label (1^4), its branch of final residency. Then, by a clockwise propeller-like motion around the fixed point (1^4), two more fixed points are created, one that moves upward and one that moves downward, away from the fixed point out of which they emerged. All three of these fixed points remain on the left-moving central branch $C^4 \left((1^4) \,\middle|\, \overline{(1^4)} \right)$ for the full baseline interval for which this sequence is central. This configuration of fixed points changes when the ζ-creation point $\zeta((2))$ of the next baseline interval of $\mathbf{B}_4$ is reached.

Interval $(\zeta((1)), \zeta((2))]$: A second period-doubling bifurcation is initiated in this very next baseline interval of $\mathbf{B}_4$ at the exact point of tangency $\widehat{\zeta}$ of the inverse graph in $P\,4$ in Chapter 7 labeled by 1.72000 of ζ (see also (5.20)). This event is initiated by a counterclockwise propeller-like motion around the fixed point $\overline{(1\ 2\ 1)}$, as this right-moving p-curve meets the 45°-line. This dynamical fixed point is, of course, the one created in the previous central interval that moved downward onto it branch $\overline{(1\ 2\ 1)}$ of permanent residency. Exactly at the point of tangency $\widehat{\zeta}$, two new fixed points emerge out of $\overline{(1\ 2\ 1)}$, one that moves upward and one that moves downward, away from the fixed point out of which they emerged. As ζ increases toward the value of $\zeta_3 = \zeta((2))$ of the next MSS root (see (5.38)), all three of these dynamical fixed points remain on the branch $\overline{(1\ 2\ 1)}$, with new ones moving away from $\overline{(1\ 2\ 1)}$ onto their respective branches of final residency, which are the positive branch $(1\ 3)$ and the conjugate branch $\overline{(1\ 3)}$ of the newly created p-curve at $\zeta_3 = \zeta((2))$.

Interval $(\zeta((2)), \infty)$: A saddle-node bifurcation at the tangency point $\widehat{\zeta}$ of the inverse graph $P\,4$ labeled by 1.98000 in Chapter 7 (see also (5.38)) initiates the creation of the last two fixed points. Here the (last) left-moving central p-curve $C_\zeta^4\left((1\ 3)\,\big|\,\overline{(1\ 3)}\right)$, which is created at the the MSS root $\zeta_3 = \zeta((2))$, simply meets the 45°-line, becoming tangent at $\widehat{\zeta}$ near the $P\,4$ inverse graph labeled by 1.98000, where the creation of the two fixed points $(1\ 3)$ and $\overline{(1\ 3)}$ occurs, and these labels are those of their respective final branches of residency. No previously created fixed points participate in this event. But, for the first time, a new event takes place: *Synchronous with the creation of the fixed points $(1\ 3)$ and $\overline{(1\ 3)}$ is the creation of the following set of non-central fixed points:*

$$\{(4),(3\ 1),(1\ 3),\overline{(1\ 3)},\overline{(1\ 1\ 2)},\overline{(2\ 2)},\overline{(2\ 1\ 1)},\overline{(3\ 1)}\,\}. \qquad (3.13)$$

This last event, a saddle-node bifurcation, in which the non-central branches with labels (3.13) are also created may come as a surprise, but they cannot be avoided: The set of all fixed points created in the inverse graph must account for all 2^4 branches of the inverse graph. This phenomenon is confirmed by the collection $P\,4$ of computer-generated inverse graphs.

It is important to realize that all adjacency properties of the branches of newly created curves are preserved in a bifurcation event. The synchrony of non-central bifurcation events with a central event is clearly very significant.

The detailed analysis of the origin and motions of fixed points for general even n can be effected along the lines presented above for $n = 4$ by reading off the results directly from the algorithmic-computer-generated inverse graphs $\mathbb{C}_n^*$ given in Chapter 5 and Chapter 7. This is not only limited in scope by the computational power to produce readable inverse graphs, but also by the power of the observer to detect very small changes in an inverse graph. Since it is already known that the inverse graphs are uniquely described by an algorithm, this direct method might even be efficient, but it necessarily will include a number of steps equal to the number of intervals in baseline $\mathbf{B}_n$, perhaps synthesized in some unifying scheme, yet to be discovered.

An alternative method is to recognize that such details as described above are more than is needed. It is better, perhaps, to recognize the existence of a unique solution, but to give the details only for some of its signature properties. This is the path followed in the remainder of this monograph.

3.6   Signature Properties of Fixed Points

3.6.1   More Vocabulary and Associated Events

1. A period-doubling bifurcation event is called a **simple event** if a given central p-curve splits in the simplest possible manner — the newly created central p-curve has one curve adjacent to it from above with positive labels and one curve adjacent to it from below with conjugate labels, where the uppermost and lowermost branch carrying the labels of the previous central p-curve. It is called a **compound event** if a given central p-curve splits into into $m > 1$ p-curves such that uppermost and lowermost branch carrying the labels of the previous central p-curve with m positive labels and m conjugate labels that are created simultaneously with the new central p-curve. An example

of this event is exhibited for in the inverse graphs $P\,8$ for $\zeta = \zeta((1))$ (slightly less than 1.62, depending on n).

2. A convenient way to describe fixed points is by giving the branches of the inverse graph G_ζ^n on which they are created and the final branches to which they evolve — **the branch of final residency.** This description is further enhanced by the definition of a p-curve.

3. A p-curve is denoted by $C_\zeta^n(\tau\,|\,\tilde{\tau}) \subset G_\zeta^n$, $\tau, \tilde{\tau} \in \mathbb{A}_n \cup \overline{\mathbb{A}}_n$, where τ and $\tilde{\tau}$ denote the upper branch and the lower branch, respectively, of two contiguous branches G_τ^n and $G_{\tilde{\tau}}^n$ that join smoothly at a single extremal point. The label $\tilde{\tau}$ is called the **complement** of τ at ζ, but is not used in this monograph.

4. The description of the motion of the fixed point defined by $p_\zeta(x) = x$ is conveniently described in terms of p-curves as follows: This fixed point belongs to the lower branch $\overline{(n)}$ of the primordial p-curve $C_\zeta^{(n)}\left((n)\,\middle|\,\overline{(n)}\right)$ for all $1/2 \leq \zeta < 1$, which corresponds to $0 \leq x < 1$. Indeed, it is the case that $x(\zeta) = 2 - \frac{1}{\zeta}$ is a fixed point of the conjugate branch of the primordial p-curve; that is,

$$\Psi_\zeta(\overline{(n)}; x(\zeta)) = x(\zeta), \text{ for all } \zeta < 1. \tag{3.14}$$

This fixed point originates on the parabola and 45°-line at $\zeta \to 0$ and as ζ increases from 0 the primordial curve becomes tangent to the 45°-line at $\zeta = 1/2$; indeed, the dynamical fixed point $x(\zeta) = 2 - \frac{1}{\zeta}$ moves **through** the permanent fixed point $(0,0)$ at $\zeta = 1/2$ on its way toward the central point $(1,1)$ of the graph. It then moves smoothly onto the upper branch (1^n) of the newly created central p-curve

$$C_\zeta^{(n)}\left((1^n)\,\middle|\,\overline{(1^n)}\right), \tag{3.15}$$

where it remains for all $\zeta > 1$, even after this central curve ($n \geq 2$) has been split apart by the creation of yet another new central p-curve. Thus, the only fixed point that remains on the branch $\overline{(n)}$ is the origin $(0, 0)$. A fixed point on the branch (n) of the inverse graph always occurs, but at the greatest MSS root $\zeta((1\;n-1)) = (\zeta((1))$ for $n > 1$, where it remains in motion on this branch for all $\zeta > \zeta((1\;n-1))$. This accounting of fixed points leads to 2^n fixed points, one on each branch of the final set of p-curves.

5. All r-cycles of the parabolic map $p_\zeta(x) = \zeta\,x(2 - x)$, $\zeta > 0$, occur for $x \in (0, 2)$ and are obtained as the the set of all positive solutions $x_i = x_i(\zeta)$ of the polynomial equation $p_\zeta^r(x) - x = 0$, where $r \in \mathbb{D}(n)$, the set of all divisors of n, denoted by the notation $n\,|\,r$. But full numerical knowledge of the set of all fixed points of the parabolic map is quite different from knowing the values of ζ at which the various fixed points make their appearance in the inverse graph. All real solutions x of $p_\zeta^r(x) = x$, $r \in \mathbb{D}(n)$, belong to the interval $(0, 2)$, and all are distinct.

But not all fixed points in a given set of such of r such points are present in the graph G_ζ^n at the same value of ζ: Each fixed point of an r-cycle has its own special ζ-value of creation — it is only for $\zeta > \zeta(1\ n-1))$, the creation point of the last branch of the inverse graph that all fixed points are present in all r-cycles. Some r-cycles are created at smaller values of ζ. Indeed, the determination of the creation ζ-value of each fixed point is quite an intricate process.

6. The guiding overview for the creation of fixed points is the following summary of proved results: In the context of the parabolic map, fixed points are the values of x that are uniquely defined by the underlying parabolic function $p_\zeta(x) = \zeta\, x(2-x)$, its iterations, and the resulting classification of fixed points into r-cycles by the divisors $r \in \mathbb{D}(n)$. But they are dynamical in the sense that each fixed point function is a smooth function of ζ with a definite point of creation beyond which it remains in the graph for all greater ζ, with a motion that carries it just past its point of creation onto a definite branch of the inverse graph, where it remains for all greater values of ζ. As the ζ-evolution continues, each individual r-cycle is filled-in with its full complement of r points, until finally for $\zeta = \zeta((1\ n-1))$, the full set of fixed point, 2^n in number, with one on each branch of the full inverse graph has found its place.

7. The precise ζ-values at which the sequences of an r-cycle appear in the inverse graph is nontrivial. This creation process may be described as follows: The branch $G_\zeta^n(\tau)$ of the graph G_ζ^n meets the line $y = x$ at a point $(\widehat{x}, \widehat{x})$ if there exists a real number $\widehat{\zeta} \in (0, \infty)$ and an $\widehat{x} = x_{\widehat{\zeta}}(\tau)$ such that the following two conditions, derived from elementary calculus, hold: The point $\widehat{\zeta}$ is a fixed point of the graph $G_{\widehat{\zeta}}^n(\tau)$ such that the graph is also tangent to the line $y = x$, as expressed by

$$\left(\widehat{x}, \Psi_{\widehat{\zeta}}(\tau; \widehat{x})\right) = (\widehat{x}, \widehat{x}), \quad \left(\widehat{x}, \Psi'_{\widehat{\zeta}}(\tau; \widehat{x})\right) = (\widehat{x}, 1), \quad \widehat{x} = x_{\widehat{\zeta}}(\tau), \quad (3.16)$$

where the prime denotes the derivative with respect to x.

8. Fixed points reside at points on a branch that intersect the 45°-line. The detailed numerical values of the creation of a fixed point are given by (3.16), which is nontrivial to effect from these relations. It entails the description of the numerical-valued smooth ζ-evolution of the x-coordinates. This process itself requires a deeper look as to how fixed points are created and evolve in ζ through saddle-node and period-doubling bifurcation events when a branch of the inverse graph G_ζ^n meets the 45°-line.

9. It is the dynamical coordinates $(\widehat{x}, \widehat{x})$ in (3.16) of the fixed points that are of interest, as well as the number of such. Their explicit analytic form is generally unknown, but they can be tracked in the set $\mathbb{C}_n^*$ of real deterministic-computer-generated graphs given in Chapter 7. The determination of the value of ζ at which all fixed points are present in the inverse graph $\mathbb{C}_n^*$ is quite restrictive. This is because the geometry inherent in how a continuous inverse branch curve, which is changing

continuously with the parameter ζ, meets the 45°-line is itself very limiting. As already noted, it is always the case that two new fixed points are created in coincidence, and then move apart with increasing ζ with a smooth motion onto their own intrinsic branch of the inverse graph G_ζ^n, where they remain for all greater values of ζ. Thus, except for values of ζ near the creation point of new fixed points, there is a one-to-one relationship between the sequences that label the branches of G_ζ^n and the set of x-coordinates that label the fixed points. The rule of assignment is: Each fixed point is uniquely labeled by its branch of final residency, and this is the label assigned.

10. Caution must be exercised in recognizing the creation of ζ-synchronous p-curves under the transformation of each central p-curve to the central p-curve for the next central interval. The process is fully deterministic in character and possesses a unique mathematical description. It is apparent that the classification of all labels of p-curves of the branches of the inverse graph at each value of ζ, together with their points of creation, is intertwined in a basic way with the concept of the cyclic permutation classification of words into equivalence classes, and, in particular, with the identification of all central self-conjugate p-curves and the basic ζ-intervals for which they are central.

Where does the above quite diverse collection of properties of fixed points leave the determination of their sequence? Though unique, no general procedure for the determination of the label of their branch of final destination has yet been given, other than by direct tracking of the ζ-evolution of the inverse graphs $\mathbb{C}_n^*$. But there is a simpler rule. Such is next described.

There is one value of ζ where the branch of final destination is known for every fixed point, namely, at $\zeta = 2$: Since every branch of the inverse graph is present at $\zeta = 2$, it must also be the case that every fixed point is present and is on its branch of final destination; namely, there is exactly one fixed point on each of the 2^n branches that constitute the full collection of branches of the inverse graph, including the conjugates. It is also the case that a fixed point reaches its branch of final destination, **and remains there,** for all ζ greater than the central branch onto which it moved just after its creation. The ζ-value of creation itself is at the exact point of tangency of the central branch. This means that the branch of final destination is on the branch that is central for the interval (ζ_t, ζ_{t+1}), which is fully known, and computable.

Section 5.2 below further develops and illustrates this property. It is a solved problem. Nonetheless, it is still useful to illustrate the many guises under which the determination of the branch of final destination manifests itself. Some of these are presented in the remainder of this Chapter 4.

There is still an important question to be answered about the approach used in this monograph to chaos theory. The branch functions $\Psi_\zeta(\alpha; x)$ are fully defined in terms of n square roots in (1.53)-(1.54), where special cases (1.17) illustrate them for $n = 4$. The ensuing theory that has been developed enforces reality conditions of a very special sort; namely, that once a branch function becomes real, it is to remain real for all greater ζ. It is these conditions that define the set $\mathbb{C}_n^*$. It may well be said that it is quite remarkable that there exists a unique pathway through the intracacies of chaos theory with such properties. But it is, of course, just this nontrivial property that makes the approach interesting, with the possibility that nontrivial, verifiable applications will ensue. It is with this spirit in mind that fixed points are visited again (see Sect. 1.2.2) to illustrate their elusive, but fundamental role and properties. The remainder of this Chapter 3 also continues along this path.

3.6.2 r-Cycles, Permutation Cycle Classes, and Words

An r-cycle has been defined already in Sect. 1.1.1 in the context of the present problem in terms of the parabolic map $p_\zeta(x) = \zeta\, x(2-x)$ by $x_{i+1} = p_\zeta(x_i)$, $i = 1, 2, \ldots, r$, subject to the condition $x_{r+1} = x_1$. It is useful to illustrate further the basic role of the dynamical fixed point $2 - \frac{1}{\zeta}$ within the context of the inverse graph G_ζ^n by this example $p_\zeta(x) = x$, which has the unique solution

$$x(\zeta) = 2 - \frac{1}{\zeta}, \quad \zeta \in (0, \infty). \tag{3.17}$$

The two branches of the primordial p-curve $C_\zeta^{(1)}((1)\,|\,\overline{(1)})$ are given by:

$$\Psi_\zeta((1); x) = 1 + \sqrt{1 - \frac{x}{\zeta}}, \quad \Psi_\zeta(\overline{(1)}; x) = 1 - \sqrt{1 - \frac{x}{\zeta}}. \tag{3.18}$$

The value of each of these functions at $x = 2 - \frac{1}{\zeta}$ gives:

$$\Psi_\zeta\left((1); 2 - \frac{1}{2}\right) = 2 - \frac{1}{\zeta}, \text{ provided } \zeta \geq 1; \tag{3.19}$$

$$\Psi_\zeta\left(\overline{(1)}; 2 - \frac{1}{\zeta}\right) = 2 - \frac{1}{\zeta}, \text{ provided } \zeta \leq 1. \tag{3.20}$$

Quite generally, it is also the case that:

The coordinate $x(\zeta) = 2 - \frac{1}{\zeta}$ is the dynamical fixed point belonging to the lower branch of the primordial p-curve $C_\zeta^n\left((n)\,\middle|\,\overline{(n)}\right)$ for all $\zeta \leq 1$. For $0 < \zeta < 1/2$, it is at negative values of $x(\zeta)$ moving toward the origin $(0,0)$ with increasing ζ; for ζ in the domain $1/2 \leq \zeta \leq 1$, it becomes exactly tangent to the 45°-line at $\zeta = 1/2$ at the origin $(0,0)$. It then moves through $(0,0)$, and continues toward the central point $(1,1)$ of the graph. It moves over at $\zeta = 1$ onto the upper branch (1^n) of the newly created central p-curve $C_\zeta^n\left((1^n)\,\middle|\,\overline{(1^n)}\right)$, where it remains for all $\zeta > 1$.

In keeping with the remarks made in the paragraph preceding the present section, it is appropriate to illustrate how the motion of the primordial p-curve initiates the entire process of creating the inverse graph for general n. For this, only the properties that relate to the primordial p-curve itself will and should be used.

The fixed point property for general n is shown directly from the recurrence relation for the conjugate branch function and the reflection relation between branch functions and their conjugates:

$$\Psi_\zeta((n); x) = 1 + \sqrt{1 - \frac{1}{\zeta}\Psi_\zeta((n-1); x)}, \text{ all } n \geq 1;$$

(3.21)

$$\Psi_\zeta(\overline{(n)}; x) = 2 - \Psi_\zeta((n)); x), \text{ all } n \geq 0.$$

A proof of these relations can be given as follows:

Proof. Each of these relations is easily proved separately by induction on n, using their validity for $n = 0, 1$, as shown above.

The tangency condition at $\zeta = 1/2$ also follows directly by induction on n from the first relation in (3.21) and its validity for (1), as follows:

$$\frac{d}{dx}\Psi_\zeta((1); x)\Big|_{x=2-\frac{1}{\zeta}} = \frac{1}{2(1-\zeta)} = 1 \text{ at } \zeta = 1/2;$$

$$\frac{d}{dx}\Psi_\zeta((n); x)\Big|_{x=2-\frac{1}{\zeta}} = \frac{1}{2\zeta\sqrt{1 - \frac{1}{\zeta}\Psi_\zeta\left((n-1); 2 - \frac{1}{\zeta}\right)}}$$

$$\times \frac{d}{dx}\Psi_\zeta((n-1); x)\Big|_{x=2-\frac{1}{\zeta}}$$

(3.22)

$$= \frac{d}{dx}\Psi_\zeta((n-1); x)\Big|_{x=2-\frac{1}{\zeta}} = 1, \text{ at } \zeta = 1/2.$$

The above relations (3.17)-(3.22) show clearly the underlying controlling role of the elusive dynamical primordial fixed point $2 - \frac{1}{\zeta}$ for all $\zeta \in (0, 1]$. But this is not yet all.

The above process can be carried forward in ζ to unveil the unfolding features of events to follow in the **successive baseline intervals.** The first creation event takes place for $\zeta_1 = \zeta((0)) = 1$, the smallest of the MSS roots. Here a family of p-curves is created **simultaneously** (synchronously) with a new central p-curve as next described:

The primordial p-curve $C_\zeta^n\left((n)\,\big|\,\overline{(n)}\right)$ is replaced at $\zeta = 1$ by the set of n p-curves with positive and negative branches labeled from top-to-bottom in the inverse graph G_ζ^n by the relations:

$$\Psi_\zeta((n - r + 1\ 1^{r-1}); x), \ r = 1, \ldots, n; \ x \in (1, \zeta_2];$$

(3.23)

$$\Psi_\zeta(\overline{(n - r + 1\ 1^{r-1})}; x), \ r = 1, \ldots, n; \ x \in (1, \zeta_2],$$

where ζ_2 is the right-most ζ-value of an interval yet to be identified. In particular, the primordial central p-curve $C_\zeta^n\left((n)\,\big|\,\overline{(n)}\right)$ has now been replaced by a new central p-curve $C_\zeta^n\left((1^n)\,\middle|\,\overline{(1^n)}\right)$ that is central in the inverse graph for all $x \in (1, \zeta_2]$. The motion of the original fixed point $x = 2 - \frac{1}{\zeta}$ during this synchronous creation of new p-curves is to move onto the upper branch $\Psi_\zeta((1^n); x)$ of the new central p-curve, where it remains for all $\zeta > 1$. It is also the case that the two branches, $\Psi_\zeta((n); x)$ and $\Psi_\zeta(\overline{(n)}; x)$, constituting the original primordial p-curve have now split apart, and all the new branches fall between these branch parts; that is, the labels ordering the branches of the new p-curves are:

$$(n-1\ 1) > \cdots > (2\ 1^{n-1}) > (1^n) >$$
$$\overline{(1^n)} > \overline{(2\ 1^{n-1})} > \cdots > \overline{(n-1\ 1)}, \ n \geq 3; \qquad (3.24)$$
$$(1\ 1) > \overline{(1\ 1)}, \ n = 2.$$

This creation event is shown in Chapter 7 at the following places for $n = 2, 3$:

$P\,2$. The creation of the central p-curve $C_\zeta^2\left((1\ 1)\,\middle|\,\overline{(1\ 1)}\right)$ is at $\zeta = 1$ (see the inverse graphs labeled by $\zeta = 1.00000$ and $\zeta = 2.10000$). No other branches are created.

$P\,3$. The creation of the central p-curve $C_\zeta^3\left((1^3)\,\middle|\,\overline{(1^3)}\right)$ is at $\zeta = 1$, and created synchronously are the branches (see $\zeta = 1.30000$ and $\zeta = 1.50000$) labeled by:

$$(2\ 1) > (1^3) > \overline{(1^3)} > \overline{(2\ 1)}. \qquad (3.25)$$

There are other branches that remain to be created, as described next in the context of arbitrary n.

These two results for $P\,2$ and $P\,3$ can be generalized to arbitrary n by using the following recurrence relation and its derivative with respect to x, evaluated at the fixed point $x = 2 - \frac{1}{\zeta}$ of $\Psi_\zeta\left((1^{n-1}); x\right)$ for $\zeta = 3/2$:

$$\Psi_\zeta\left((1^n); x\right) = 1 + \sqrt{1 - \frac{1}{\zeta}\Psi_\zeta\left((1^{n-1}); x\right)};$$

$$\frac{d}{dx}\Psi_\zeta\left((1^n); x\right)\bigg|_{x=2-\frac{1}{\zeta}} = -\frac{1}{2\zeta\sqrt{1 - \frac{1}{\zeta}\Psi_\zeta\left((1^{n-1}); 2 - \frac{1}{\zeta}\right)}}$$
$$\times \frac{d}{dx}\Psi_\zeta\left((1^{n-1}); x\right)\bigg|_{x=2-\frac{1}{\zeta}} \qquad (3.26)$$
$$= \frac{1}{-2(\zeta-1)}\frac{d}{dx}\Psi_\zeta\left((1^{n-1}); x\right)\bigg|_{x=2-\frac{1}{\zeta}};$$

$$\frac{d}{dx}\Psi_\zeta\left((1^n); x\right)\bigg|_{x=\frac{4}{3}} = -\frac{d}{dx}\Psi_\zeta\left((1^{n-1}); x\right)\bigg|_{x=\frac{4}{3}}. \qquad (3.27)$$

Relation (3.27) now gives the desired result, again by induction on n and its validity for $n > 3$:

$$\frac{d}{dx}\Psi_\zeta\left((1^{n-1}); x\right)\bigg|_{x=\frac{4}{3}} = 1, \text{ for } n-1 \text{ even implies}$$

$$\frac{d}{dx}\Psi_\zeta\left((1^n); x\right)\bigg|_{x=\frac{4}{3}} = -1 \text{ for } n \text{ odd, and conversely, all } n > 3. \text{ (3.28)}$$

Notice that the central p-curve $C_\zeta^{(n)}\left((1^n)\,\big|\,\overline{(1^n)}\right)$, $\zeta \in (1, \zeta_2]$, is a left-moving curve. For $\zeta = 3/2$ and odd n, the branch (1^n) is perpendicular to the $45°$-line; for $\zeta = 3/2$ and even n, it is tangent to the $45°$-line. This is true for all $n > 3$, but fails to be the case for $P\,2$ and $P\,3$. This shows nicely that caution must always be exercised in anticipating events in the inverse graph.

The results obtained above in the inverse graphs $P\,2, P\,3$, but now including the properties (3.26)-(3.28), give a universal behavior for all n of the the set of inverse graphs $P\,n$. In particular, this shows up for the first two intervals $\zeta \in (0, 1]$ and $\zeta \in (1, \zeta_2)$, where ζ_2 is the MSS root adjacent to $\zeta = 1$, that is, $1 < \zeta_2$ in the set MSS of all MSS roots, and $\zeta = (1+\sqrt{5})/2$ for $n = 2$. The central curves are $C_\zeta^n\left((n)\,\big|\,\overline{(n)}\right)$ and $C_\zeta^n\left((1^n)\,\big|\,\overline{(1^n)}\right)$, and the fixed points are $(0,0)$ at the origin, and the dynamical fixed point $x(\zeta) = 2 - \frac{1}{\zeta}$, which emerges out of the origin $(0,0)$ at $\zeta = 1/2$ and moves onto its permanent branch (1^n) of residency at $\zeta > 3/2$. The parameter value $\zeta = 3/2$ is itself the exact creation point of two new fixed points that initiate the following events:

For odd n, only saddle-node bifurcation events occur, exactly once in successive baseline intervals, which are exhausted after all (a unique number) baseline intervals have been exhausted.

For even n, a series of period-doubling bifurcation events occur, one in each successive baseline interval, until no more are allowed by the constraints on the total number 2^n of fixed points, now including the saddle-node bifurcations, that can appear in the inverse graph $P\,n$ for all $\zeta \in (0, \infty)$. *The number of each type of bifurcation is unique.*

It is useful next to show how the combinatorial properties of words enter into this analysis and intertwine with r-cycles.

The labels of the branches of the inverse graph can be partitioned into cyclic permutation equivalence classes in a purely combinatorial fashion that is fully divorced from the graph problem itself. This is because there is a one-to-one map between branch labels and the set of 2^n words on the two letters R and L. Thus, a first word is selected and all cyclic permutations effected and placed in an equivalence class. A second word, not in the first class, is then selected, and the cyclic permutations effected to obtain a second equivalence class. This process is repeated until all 2^n words are taken into account. This procedure not only gives a partition of the set of 2^n words into equivalence classes, it also contains implicitly the number of such equivalence classes. For the problem at hand, these equivalence classes are next mapped back into corresponding sets of positive integers and their conjugates, thus obtaining the labels of all branches of the inverse graph, now classified into equivalence classes by the cyclic permutations of their corresponding words.

The sequence of steps just described can be detailed as follows: Define

$$\mathbf{W}_n(\tau) = \{w(\tau') \in \mathbb{W}_n \,|\, w(\tau') \equiv w(\tau)\}, \tag{3.29}$$

where τ runs over all distinct sequences as required to enumerate all such equivalence classes. Once this partitioning has been effected, it is always possible to enumerate each equivalence class by choosing as class representative the greatest positive sequence $\alpha_{\max}$ contained in each set, except for the single case of $\overline{(n)} \mapsto L^n$:

$$\mathbf{W}_n(\alpha_{\max}) = \{w(\tau) \in \mathbb{W}_n \,|\, w(\tau) \equiv w(\alpha_{\max})\}. \tag{3.30}$$

Correspondingly, the sets of equivalent sequences are defined:

$$\mathbf{C}_n(\alpha_{\max}) = \{\tau \in \mathbb{W}_n \,|\, w(\tau) \equiv w(\alpha_{\max})\}. \tag{3.31}$$

This result illustrates a structural aspect of cycles that cannot be emphasized too strongly in the present work:

The partitioning of the labels of the branches of the inverse graph into equivalence classes is a purely combinatorial problem fully divorced from the graph problem itself because there is a one-to-one map between these labels and the set of 2^n words on the two letters R and L. Nonetheless, it is the case that further details on the motion of fixed points can be obtained from the inverse graph itself.

This Chapter 3 now continues with the description of still another collection of patterns coming from a completely different source that may be used to describe the properties of the inverse graph $\mathbb{C}_n^*$.

3.7   Young Hook Tableaux and Gelfand-Tsetlin Patterns

The Young hook standard tableaux that occur here are hardly sufficient for showing the rich relationships between Young tableaux and Gelfand-Tsetlin patterns. Much of this can be found in Ref. [53], where the same notations and nomenclature are used. A more abstract and general description of hook tableaux can be found in Stanley [54]. It is the case that the number of hook tableaux for general n is 2^n, as demonstrated in (3.33) below for $n = 1, 2, 3$.

The enumeration of the labels of the inverse graph G_ζ^n by standard hook tableaux was noted by Stein [14] and again by Bivins, *et al* [15]. Gelfand-Tsetlin patterns give a one-to-one presentation of such standard hook tableaux, as will be explained below. It is, perhaps, unexpected that these combinatorial objects should occur. This occurrence clearly places the present subject within the purview of combinatorics. It is shown in this section how this takes place. It is an important nontrivial result.

A hook tableau has the shape $(n - k \; k)$, each $k = 0, 1, 2, \ldots, n - 1$:

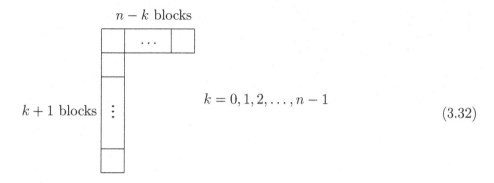

$k = 0, 1, 2, \ldots, n - 1$

(3.32)

The **length** of each hook tableau (number of blocks) is n, there being $n - k$ blocks in row 1 (the top row), and $k + 1$ such blocks in the single column), with one shared block in row 1 and column 1, as shown.

The standard hook tableau in (3.32) is to be filled-in with the integers $1, 2, \ldots, n$, one integer in each block, such that the collection of integers appearing in row 1 and in column 1 are each strictly increasing. The **content** or **weight** of the hook tableau is (1^n), while its **shape** is the **partition** $(n - k \; 1^k)$.

Examples. $n = 1, 2, 3$:

$$
\boxed{1}
$$

$$
\begin{array}{cc} \boxed{1} & \boxed{2} \end{array} \qquad \begin{array}{c} \boxed{1} \\ \boxed{2} \end{array}
$$

(3.33)

$$
\begin{array}{ccc} \boxed{1} & \boxed{2} & \boxed{3} \end{array} \quad \begin{array}{cc} \boxed{1} & \boxed{2} \\ \boxed{3} \end{array} \quad \begin{array}{cc} \boxed{1} & \boxed{3} \\ \boxed{2} \end{array} \quad \begin{array}{c} \boxed{1} \\ \boxed{2} \\ \boxed{3} \end{array}
$$

The integers in the filled-in standard tableaux enumerated by (3.33) do not give directly the set of labels of the branches that occur in the Creation Tables $\mathbb{T}_n, n = 1, 2, 3, 4$, nor do the integers in the filled-in general hook tableaux in (3.33) give directly the set of labels of the branches that occur in the Creation Table $\mathbb{T}_n$. There is no reason that this should be the case. But the number 2^n of each agrees. To make the connection between the two sets of labels, it is convenient to use GT patterns, which have a definition for arbitrary n and have many applications in physics and mathematics, as is also true for Young tableaux. The special application to hook tableaux is then made for the present problem.

Let $\lambda = (\lambda_1 \geq \lambda_2 \geq \lambda_n \geq 0)$ denote a partition consisting of n non-negative integers, with 0 counted as a part. The general GT pattern for arbitrary n consists of a partition λ together with $n(n-1)/2$ nonnegative integers arranged in a triangular array denoted by $\binom{\lambda}{m}$ and of the following form:

$$
\binom{\lambda}{m}
$$

$$
= \quad
\begin{matrix}
\lambda_1 & \lambda_2 & \cdots & \lambda_{n-1} & \lambda_n \\
 & m_{1\,n-1} & m_{2\,n-1} & \cdots & m_{n-1\,n-1} \\
 & & \vdots & & \vdots \\
 & & m_{1\,2} & m_{2\,2} & \\
 & & & m_{1\,1} &
\end{matrix}
\tag{3.34}
$$

The entries $m_{i\,j}$ are to satisfy conditions known as the "betweenness relations," which may be stated as follows in terms of each basic triangular array in (3.34) of the form:

$$
\begin{pmatrix}
m_{i\,j} & & m_{i+1\,j} \\
 & m_{i\,j-1} &
\end{pmatrix}
$$

$$
m_{i\,j} \geq m_{i\,j-1} \geq m_{i+1\,j}, \quad j = 2, 3, \ldots, n;
\tag{3.35}
$$

$$
\lambda_i = m_{i\,n}; \quad i = 1, 2, \ldots, n.
$$

The placement of symbols in (3.34)-(3.35) denotes that the numerical value of the lower symbol, which is placed between the upper two symbols, can be any value between and including the upper two symbols. Thus, starting with a given partition λ, the full set of patterns is uniquely prescribed. The number of such GT patterns is given by the well-known Weyl dimension formula (see Ref. [53]), which is not needed here, and is not stated. There is still another important property of general GT patterns that needs to be defined. It is called the weight of the GT pattern $\binom{\lambda}{m}$:

$$
W\binom{\lambda}{m} = (W_1, W_2, \ldots, W_n); \quad W_i = \text{sum of integers in row } i \text{ of } \binom{\lambda}{m}
$$

$$
- \text{ sum of integers in row } i-1 \text{ of } \binom{\lambda}{m},
$$

$$
i = 2, 3, \ldots, n; \quad W_1 = m_{1\,1}.
\tag{3.36}
$$

The general pattern corresponding to the special hook partition $\lambda =$

$(n - k \ \ 1^k \ \ 0^{n-k-1})$ is given by

$$
\begin{pmatrix}
n - k & 1^k & 0^{n-k-1} \\
& \vdots & \vdots & \\
& m_{1\,2} & m_{2\,2} & \\
& & 1 &
\end{pmatrix}
\qquad \text{weight} = (1^n),
\tag{3.37}
$$

where, for each n and each $k = 0, 1, \ldots, n - 1$, the pattern is to be filled-in in all possible ways that give the weight (1^n), where the weight is defined by (3.36).

It is appropriate to note again that while nothing close to the richness of structure of the general GT patterns is needed for this special case, it is still important to know about the general theory from which hook patterns emerge as given above.

The number of hook patterns and the number of positive labels in the general Creation Table $\mathbb{T}_n$, and their conjugates, is 2^n, as noted above, but **the weight of each of the hook patterns is** (1^n); that is, all hook patterns have the **same weight.** Thus, the weight is not the parameter needed to obtain a one-to-one correspondence with the hook patterns. An appropriate order relation on the set of 2^{n-1} GT patterns corresponding to the hook tableaux (3.37) is obtained by associating the following sequence of length $\binom{n}{2}$ to the GT pattern (3.37):

$$
(\text{row } n \quad \text{row } (n - 1) \quad \cdots \quad \text{row } 2 \quad \text{row } 1).
\tag{3.38}
$$

In this relation, the n rows of the GT pattern (3.37) are placed in a **single row,** where the rows of the GT pattern (3.37) are read from top-to-bottom and left-to-right in (3.38), as displayed.

It is instructive to look at the example of the GT patterns for a simple case, say, $n = 3$, where k can be $k = 0, 1, 2$. Thus, the partition can be $(3\ 0\ 0), (2\ 1\ 0), (1\ 1\ 1)$, and the corresponding GT patterns are obtained by filling in the 3-rowed triangular pattern in all ways that give the weight $(1\ 1\ 1)$. The patterns so obtained are the following four, where the sequence (3.38) is placed after the GT pattern for ease of comparison:

$$
\begin{pmatrix} 3 & 0 & 0 \\ & 2 & 0 & \\ & & 1 & \end{pmatrix}, \quad (3\ 0\ 0\ 2\ 0\ 1); \quad \begin{pmatrix} 2 & 1 & 0 \\ & 2 & 0 & \\ & & 1 & \end{pmatrix}, \quad (2\ 1\ 0\ 2\ 0\ 1);
$$

$$
\tag{3.39}
$$

$$
\begin{pmatrix} 2 & 1 & 0 \\ & 1 & 1 & \\ & & 1 & \end{pmatrix}, \quad (2\ 1\ 0\ 1\ 1\ 1); \quad \begin{pmatrix} 1 & 1 & 1 \\ & 1 & 1 & \\ & & 1 & \end{pmatrix}, \quad (1\ 1\ 1\ 1\ 1\ 1).
$$

Standard page ordering on the sequences (3.38) now gives the following order relations on the GT patterns in (3.39). (Reverse-lexicographic order gives the same order relation as in (3.40).) They are equivalent order relations:

$$\left(\begin{smallmatrix} 3 & 0 & 0 \\ & 2 & 0 \\ & & 1 \end{smallmatrix} \right) > \left(\begin{smallmatrix} 2 & 1 & 0 \\ & 2 & 0 \\ & & 1 \end{smallmatrix} \right) > \left(\begin{smallmatrix} 2 & 1 & 0 \\ & 1 & 1 \\ & & 1 \end{smallmatrix} \right) > \left(\begin{smallmatrix} 1 & 1 & 1 \\ & 1 & 1 \\ & & 1 \end{smallmatrix} \right). \tag{3.40}$$

Here, the simple page ordering on the sequences (3.38) has been transferred to the GT patterns themselves, since they are one-to-one.

Relation (3.40) generalizes in the obvious way to arbitrary n. Indeed, these results mean that an entirely different analysis of the entries in the Creation Table $\mathbb{T}_n$ than presented in this monograph can be given **based on GT patterns alone.** This must be the case because of the existence of the one-to-one correspondence of GT patterns of the form (3.37) to the labels assigned to the collection of inverse graphs by the method of reverse-lexicography.

Every property of the Creation Table $\mathbb{T}_n$ can be expressed as a property of the corresponding GT pattern (3.37). This has not been carried out here.

Chapter 4

The (1+1)-Dimensional Nonlinear Universe

4.1 The Parabolic Map

This short chapter from which this monograph derives its title is interpretive and conjectural. It is based on the material developed in Chapters 1-3 on the many algorithmic-computer-generated inverse graphs $\mathbb{C}_n^*$ given in Chapter 7. The chapter title quite aptly describes the resulting "Theory of Everything" that ensues. The interpretive and speculative part refers to such systems, including General Relativity.

What makes the present approach distinctive in its application to general relativity is that the mathematics underlying the approach is based on function composition, which is inequivalent to the differential geometry of Einstein's Relativity.

The properties that are unveiled in the present approach are also rich in structure and broad in their applications. It is a remarkable fact that the branches of the inverse graph become real at a certain critical point in a single parameter and stay real for all greater values of the parameter. Moreover, the theory is enriched by connections with abstract combinatorial concepts, such as Young tableaux, Gelfand-Tsetlin patterns, and the theory of words on two letters. What is the most unexpected is the possibility of connecting these mathematical structures to that of the real Universe itself.

The idea that suggests itself originates with Einstein, who is reputed (Pais [41]) to have said:

One of the most remarkable things about the Universe is that it is comprehensible.

One cannot help but wonder if Einstein considered the idea:

One of the most remarkable things about the Universe is that it is incomprehensible.

Remarks. It seems to be quite difficult to find a primary source for statements attributed to famous scientists. The attribution to Einstein above
appears to be consistent with various statements made in Pais [41]. In any
case, placing the two statements in direct apposition serves quite well our
purpose here, which is to show the relation of the two statements to the algorithmic approach to the function composition and the nonlinear properties
that ensue. In the usual approach to general relativity, it is the generalization of differential equations to differential geometry and the continuity
of space-time that yields the great insights that allows the real Universe to
evolve.

Definitional terms and speculative interpretations that are consistent with
the viewpoint of **Chaos Theory** as a **Complex Adaptive System**, as
presented in this monograph, are next given:

1. **Objects:** In this algorithmic approach to the behavior of complex
objects, it is the "objects" that are undefined. Their definition depends on
the application.

2. **Equations of motion:** There are no equations of motion as such —
the "motion of objects" is fully governed by the shape of the curves during
the ζ-evolution, which *creates its own nonlinear space.*

3. **Shape of the inverse space:** The shape of the inverse space at
each value of ζ is defined to be the set of points belonging to the inverse
graph at the selected value of ζ. It is always a set of continuous points joined
smoothly at all extremal points. Many examples are shown in the collection
of computer-generated inverse graphs given in Chapter 7.

4. **One-dimensional space:** The space in which objects move is one-
dimensional. The forty-five degree line is not part of the space, nor are the
vertical lines at $x = 0$ and $x = 2$, except that the line segment $x = 0, y \in [0, 1]$
at $\zeta = 0$ constitutes the entire shape of the graph. The redundancy of lines
is included simply to help visualize the structure of the one-dimensional
space along which objects move. To move between two points belonging to
the one-dimensional space it is essential that the motion of any object be
confined to the points defining the shape of the curve.

5. **Creation of objects:** The entities called objects are created in pairs
which are called objects and anti-objects. Objects and anti-objects have the
following properties:

i. Each pair is created at an MSS root, which is characteristic of the
object. In all, at $\zeta = 2$, there are 2^{n-1} such pairs in the inverse graphs with
a lesser number for $0 \leq \zeta \leq (1\ n - 1), n \leq 2$. There are no pairs created
for $\zeta > 2$, although the graph continues to undergo changes as it evolves
continuously.

ii. Objects are always created, never annihilated, and once created are
dynamical objects that move apart, always along the shape of the curve.

iii. The environment of an object at a given ζ-value is the collection of
objects present in the inverse graph at that value of ζ. The environment is
a dynamical property.

iv. There is a special class of objects called **central objects** that are created on a central curve. Central curves are those whose extremal point belongs to the central line $y = 1$. Central curves oscillate about the central point $(1, 1)$ of the graph with a variable amplitude that depends on the pair of MSS roots of the baseline interval of the central curve, this feature holding for all $0 \leq \zeta \leq 2$. But, for $\zeta > 2$, the last central curve created, which has positive label $(1\ n - 1)$, $n \geq 2$ and conjugate label $\overline{(1\ n - 1)}$ moves leftward for all $\zeta > 2$ and moves completely (is ejected) out of the graph at $\zeta = 0$, and continues its leftward motion for all greater ζ.

6. **Black holes as objects:** Black holes can be taken as objects. There are two types of black holes created by the two types of bifurcations.

7. **Matter and antimatter as objects:** Matter (positive sequences) and antimatter (conjugate sequences) can be taken as objects. Note then from 5(iv) above that matter and antimatter share but a single point throughout their existence, that is, for all $\zeta > 0$. This point belongs to the central line $y = 1$: It is the point described in (iv) that undergoes the oscillatory motion described there. The asymmetry of matter versus antimatter is a consequence of the forty-five degree line determining the points of creation of all objects, even though that line is not part of the shape of the inverse graph. The "skewness", however, remains in the creation of objects.

8. **Central curves control all:** The analogues of the primordial curve curve $\mathcal{C}((n); \overline{(n)})$ are the central curves $\mathcal{C}(c_n(t); \overline{c_n(t)})$, $t = 1, 2, \ldots, |L_n| + 1$: At $t = 1$, new curves are created, then again at $t = 2, \ldots$, then again at $t = |L_n| + 1$, and lastly at $t = |L_n| + 1$. Thus, the same process is repeated over and over, with new curves being interjected into the graph at each creation point.

9. **It is the microscopic world that is incomprehensible:** The quantum world of electrons, protons, etc. can never be rationally explained.

10. **Combinatorial structure:** Based abstractly on special classes of words on two letters, called R and L with which there is a one-to-one relation, which then also maps to a class of binary numbers. All information is therein contained. The Universe "knows" how to count.

11. **Collapse and regeneration:**

(i). For $\zeta > 2$, the branches in the inverse graph appear to undergo an intricate evolution that entails a merger of families of adjacent branches to the neighborhood of a characteristic horizontal line. See the computer-generated inverse graphs in the collection $P8$ labeled by $\zeta = 2.00000$ and $\zeta = 2.20000$, as well as the labellings of graphs in (5.35)-(5.42). The families of sequences in question are the ordered sets defined for $k = 1, 2, \ldots, n - 1$ by $\mathcal{F}_k^n = \{(k + 1\ 1^{n-k-1}), \ldots, (k\ 1^{n-k})\}^{ord}\rceil$, and for $k = 0$ by $\mathcal{F}_0^n = \{(1^n), \ldots, (1\ n - 1)\}^{ord}$, where it is recalled that the operation $\rceil$ acts from the left on a set of sequences and removes the right-most sequence. These sequences are shown for $n = 8$ in the list (5.42), as interpreted for the two computer-generated graphs referred to above. The inverse graph labeled by $\zeta = 2.0000$ and the ones preceding this value show that all 2^n inverse graphs are present at $\zeta = 2$. But a new phenomenon appears as ζ increases beyond $\zeta = 2$. The p-curves coalesce to the neighborhood of eight band-like structures, although the fact that the computer-generated branches are not resolved leaves uncertainty. It is, however, a fact that all 2^{n-1} positive sequences must be present and their order preserved. But the dynamical process by which this take place as ζ increases requires further theoretical developments and computer-verified calculations. Of course, the conjugate sequences follow a symmetrical process.

(ii). The most surprising of all, perhaps, is the behavior of the branches of the inverse graph for negative values of the parameter ζ (see the set of computer-calculated inverse graphs for $P\,3$ and $P\,8$). For large negative values of ζ, the branches of the inverse graph appear to be distributed in a different way for n odd and n even. But this is highly speculative because the theoretical and computational background has not yet been done to show how the motions of the branches in the domain of x outside the domain $[0,2]$. It is left as an open problem to carry out such calculations, since present resources are not available here for such.

(iii). It is worth noting here that the so-called set of universal sequences given by $\{(n),(n-1\;1),(n-2\;1\;1),\ldots,(1^n)\}^{ord}$ most certainly have a major structural role. Also, all changes in shape of the inverse graph as ζ increases from lesser to greater negative values motions must be such as to allow a smooth ζ-evolution into the inverse graph for the interval $(x,\zeta)\;\in\;[0,2]$ of the inverse graph, as presented in this monograph. It is important to understand that the *shape of the inverse graph outside the interval* $[0,2]$ *is fully determined:* It is simply that sufficiently many calculations have not yet been done to to determine it. There is still much to be learned.

12. **Universality:** This refers to the concave downward property of the parabola and the fact that all curves possessing this property exhibit similar shapes under function composition. There are many presentations elsewhere of this property and its meaning. It is not discussed in this monograph, although it could be important.

13. **Quantum harmonic oscillator states:** The $2n-1$ positive sequence labels can be presented by filled-in Young standard tableaux known as **hook tableaux.** These tableaux can also be realized by what are known as Gelfand-Tsetlin patterns of integers. The one-to-one relation between these sets is well-known and is presented in great detail in Ref. [53] with many references to the published literature. This is presented in Sect. 3.5 in the context of the present problem. What is important here is that the Gelfand-Tsetlin patterns can be realized explicitly in terms of a collection of a class of isotropic quantum harmonic oscillators. Thus, such oscillators occur naturally: They are created at the MSS roots, and their number in one-to-one with the creation of new sequences in the present algorithmic approach to complex systems. The geometry of the ζ-evolution of the inverse graph admits naturally a quantum-mechanical classification of different types of fundamental particles, but here its meaning is left open.

14. **Supernatural elements:** A realistic model of the Universe should admit the question: Do supernatural influences exist in this model? A contradictory answer is taken to mean that the existence of such influences can neither be proved or disproved **within the framework of the model.**The computer-generated inverse graph developed in this monograph is a model

of a complex system that admits the interpretation of Scripture given above. As applied to the existence of God, where the term "God" is used in the generic sense of any reasonable influence, the biblical notion that God is both within us and with us[1] can be interpreted as a contradiction, that is, the existence of God can neither be proved nor disproved within the framework of the subject of this monograph.

4.2 Complex Adaptive Systems

The algorithmic-computer-generated approach to complex systems can also be applied to the first statement attributed above to Einstein that *one of the most remarkable things about the Universe is that it is comprehensible.* This application will lead to a different model than the one presented above in this chapter, as well as the one of Einstein's General Relativity. This is true because it is based on the fundamental role of the mathematical operation of function composition, not on the model of differential equations (Maxwell) carried out by Einstein in formulating his General Relativity in terms of differential geometry and topological spaces. The Incomprehensibility of the Universe model given above and its tantalizing possibilities has been preferred here. The "incomprehensibility of the quantum world" statement is, perhaps, too severe; it must be remembered that possible interpretations of a model are only suggestive of properties of the real Universe and need not be realized. It is also the case that the results above can be recast in terms of the first statement above that the universe is comprehensible. Since it is function composition that is involved, this model of general relativity is still distinct from Einstein's.

---

[1] See, for example, "Where is the Kingdom of God? Is it in your Heart?" by guest contributor Mark Droberts to Patheos Newsletter May 2011, Vol. 19.

Chapter 5

The Creation Table

The Creation Table $\mathbb{T}_n$ is a collection of columns in which the elements that stand in the same column as the central sequence $c_n(t)$ are precisely those created synchronously with ζ_t at the MSS root ζ_t that defines the interval $(\zeta_t, \zeta_{t+1}]$ for which $c_n(t)$ is the central sequence. The formatting of such a table presents problems even for small n because each row contains only one entry, there being a number of rows in the table equal to the number of baseline intervals in $\mathbf{B}_n$, and the Table contains 2^n distinct labels in all, including conjugate labels. Nonetheless, the full Table is exactly known from the deterministic-computer-generated inverse graphs $\mathbb{C}_n^*$ given earlier. These Creation Tables are repeated here in this Chapter for convenience of reference to the computer-generated inverse graphs $\mathbb{C}_n^*$ in Chapter 7:

The starting place is for $n = 1$, where the baseline $\mathbf{B}_1$ consists of two intervals as depicted in the following diagram:

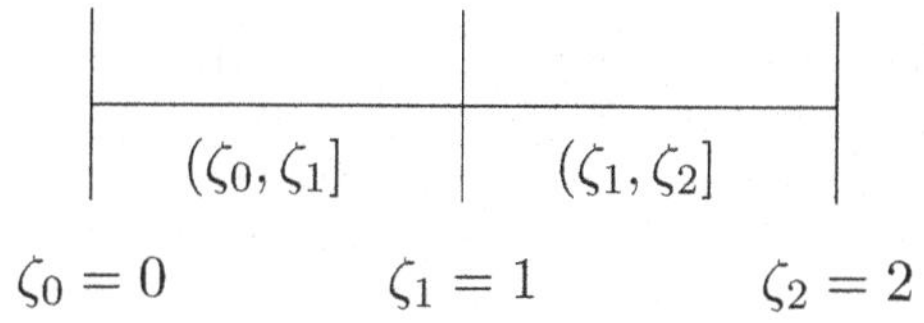

$$
\begin{array}{ccc}
\vert & \vert & \vert \\
(\zeta_0, \zeta_1] & (\zeta_1, \zeta_2] & \\
\zeta_0 = 0 \quad & \zeta_1 = 1 \quad & \zeta_2 = 2
\end{array}
\tag{5.1}
$$

Here $\zeta_0 = 0$ and ζ_2 mark the endpoints of the baseline and are not MSS roots. The information on the creation sequences can be presented in the following way:

The new creation sequences present at each central interval for $n = 1$:

$$
\begin{aligned}
\mathbb{C}_1^*(\zeta_0) &= \{(1)\}, \text{ for } \zeta \in (\zeta_0, \zeta_1]; \\
\mathbb{C}_1^*(\zeta_1) &= \{\overline{(1)}\}, \text{ for } \zeta \in (\zeta_1, \zeta_2].
\end{aligned}
\tag{5.2}
$$

The notations $\mathbb{C}_1^*(\zeta_0)$ and $\mathbb{C}_1^*(\zeta_1)$ designate that the respective sequences (1) and $\overline{(1)}$ are created at $\zeta_0 = 0$ and $\zeta_1 = 1$ (the asterisk denotes **created**).

Baseline $\mathbf{B}_2$ consists of three intervals:

$$\zeta_0 = 0 \qquad\qquad \zeta_1 = 1 \quad \zeta_2 = (1 + \sqrt{5})/2 \qquad \zeta_3 = 3$$

(5.3)

This diagram gives the new positive creation sequences present at each central interval for $n = 2$:

$$\begin{aligned}
\mathbb{C}_2^*(\zeta_0) &= \{(2)\}, \text{ for } \zeta \in (\zeta_0, \zeta_1]; \\
\mathbb{C}_2^*(\zeta_1) &= \{(1\ 1)\}, \text{ for } \zeta \in (\zeta_1, \zeta_2]; \\
\mathbb{C}_2^*(\zeta_2) &= \phi, \text{ for } \zeta \in (\zeta_2, \zeta_3].
\end{aligned}$$

(5.4)

The notations $\mathbb{C}_2^*(\zeta_0)$, $\mathbb{C}_2^*(\zeta_1)$, designate that the positive sequences (2), $(1\ 1)$, and the empty sequence $\phi =$ no sequence are created at $\zeta_0 = 0$, $\zeta_1 = 1$, and $\zeta_2 = (1 + \sqrt{5})/2$.

Baseline $\mathbf{B}_n$ consists of q_n intervals:

$$\zeta_0 = 0 \qquad\qquad \zeta_1 = 1 \qquad\qquad \zeta_2 \qquad\qquad \zeta_{q_{n-1}} \qquad \zeta_{q_n} = q_n$$

(5.5)

This diagram (5.5) only gives the new positive creation sequences present at the first two intervals in baseline $\mathbf{B}_n$ where ζ_2 is the sequence for the MSS root adjacent to the next greater MSS root in the set $\mathbb{MSS}$ of all MSS roots, and, in addition, the so-called Universal Sequences, denoted $\mathbb{U}(\zeta_1, \zeta_2)$, are present: They are created at $\zeta = 1$ synchronously with the central sequence at the MSS root $\zeta((1)) = 1$, and are n in number, counting the central sequence. They are Universal in the sense that all sequences in the set can be given for arbitrary n, once a suitable notation for them has been given.

A compact expression for the set of universal sequence $\mathbb{U}_n(\zeta_1, \zeta_2)$, as well as other relevant sequences that arise, can be given in terms of a pair of operators, denoted $\lceil$ and $\rceil$. The operator $\lceil$ acts from the left on an arbitrary ordered set by removing the left-most element from the set, while $\rceil$ acts from the right by removing the right-most element. Thus, for an arbitrary fully ordered set $X = \{x_1, x_2, \ldots, x_t\}$ with $x_1 > x_2 > \cdots > x_t$, the actions of $\lceil$ and $\rceil$, respectively, are defined by

$$\lceil \{x_1, x_2, \ldots, x_t\} = \{x_2, \ldots, x_t\}; \tag{5.6}$$

$$\{x_1, x_2, \ldots, x_t\} \rceil = \{x_1, \ldots, x_{t-1}\}. \tag{5.7}$$

The notations $\lceil^{(r)}$ and $\rceil^{(s)}$ designate repeated applications of the respective operators a number of times given by $r = 0, 1, \ldots, t$ and $s = 0, 1, \ldots, t$. The action of these operators on the empty sequence ϕ is defined by

$$\lceil^{(r)} \phi = \phi \rceil^{(s)} = \phi. \tag{5.8}$$

It is also to be noted that the ordering of sequences in the transformed subset by either of the operators $\lceil^{(r)}$ or $\rceil^{(s)}$ is the same as that in the original set. The above features of the operators $\lceil$ and $\rceil$ are already present in the simplest example for $n = 2$, as illustrated by:

$$\lceil^{(0)} \{(2), (1 \ \ 1)\}^{ord} \rceil^{(0)} = \{(2), (1 \ \ 1)\}^{ord};$$

$$\lceil^{(1)} \{(2), (1 \ \ 1)\}^{ord} \rceil^{(0)} = \{(1 \ \ 1)\};$$

$$\lceil^{(2)} \{(2), (1 \ \ 1)\}^{ord} \rceil^{(0)} = \phi;$$

$$\lceil^{(0)} \{(2), (1 \ \ 1)\}^{ord} \rceil^{(1)} = \{(2)\};$$

$$\lceil^{(1)} \{(2), (1 \ \ 1)\}^{ord} \rceil^{(1)} = \phi; \tag{5.9}$$

$$\lceil^{(2)} \{(2), (1 \ \ 1)\}^{ord} \rceil^{(1)} = \phi;$$

$$\lceil^{(0)} \{(2), (1 \ \ 1)\}^{ord} \rceil^{(2)} = \phi;$$

$$\lceil^{(1)} \{(2), (1 \ \ 1)\}^{ord} \rceil^{(2)} = \phi;$$

$$\lceil^{(2)} \{(2), (1 \ \ 1)\}^{ord} \rceil^{(2)} = \phi.$$

The purpose of the $\lceil$ and $\rceil$ operators is very simple: They serve to identify any subsequence of the original ordered sequence, while preserving the order, and exhausting the finite set of all elements by producing the empty sequence. Thus, in the set of 2^{n-1} positive sequences in the totally ordered set $\{(n), (n - 1 \ \ 1), \ldots, (n - 1 \ \ 1)\}^{ord}$ a unique value of the pair $(r, s), r = 0, 1, \ldots, q_n$ and $s = 0, 1, \ldots, q_n$ is assigned to every nonempty sequence. But this is not yet sufficient to identify for general n the interval in baseline $\mathbf{B}_n$ to which a newly created sequence is assigned. It is also the case that the explosive growth of relations (5.9) prohibits explicit enumeration, already for quite small n. But there is a method of effecting the desired identification that goes as follows: The branches to which every fixed point belongs are known at $\zeta = 2$: They are just the 2^n branches labeled from top-to-bottom of $\mathbb{T}_n \cup \overline{\overline{\mathbb{T}}}_n$.

The following procedure then effects the desired identification of the baseline intervals of $\mathbf{B}_n$: Remove from the full set at $\zeta = 2$ exactly the set of sequences that label the branches not yet created, which are known, since the labels of all branches created up to each values of ζ are known. This must give the fixed points in the set $\mathbb{C}_n$. Indeed, this process can be carried out by hand for relatively small values of n. Despite this structural simplicity, it is still useful to implement it in more detail, since this illustrates in more detail unsuspected properties. The analysis continues along these lines.

5.1   The Creation Intervals

The distribution of sequences into their creation intervals is unique as determined by the deterministic-computer-generated inverse graphs $\mathbb{C}_n^*$. A method that gives the columns into which the sequences in baseline $\mathbf{B}_n$ are placed can be proved as follows: *Proof by recursion.* The recursive construction begins with (5.2) for $n = 1$ and follows the usual method of assuming the result for $\mathbb{T}_{n-1}$ and "lifting" the result to $\mathbb{T}_n$. The result for $n = 2$ given by (5.4) already follows from the result for $n = 1$ by adding 1 to the first position in the sequence (1) to obtain (2), and then adjoining 1 to the left end of the sequence (1) to obtain the sequence (1 1). There is one subtlety in the general transformation transformation $\mathbb{T}_{n-1} \longrightarrow \mathbb{T}_n$ that does not show up in the simple transformation $\mathbb{T}_1 \longrightarrow \mathbb{T}_2$; it is contained in statement (2.54), which can be restated in the following form, which assumes all central sequences in baseline $\mathbf{B}_n$ for arbitrary n are known, as already proved: Let $\alpha \in \mathbb{T}_{n-1}$. Then, the sequence $(1\ \alpha) \in \mathbb{T}_n$ is either a central sequence or a non-central sequence; if central, then $(1\ \alpha)$ goes to the same column in both baseline $\mathbf{B}_{n-1}$ and baseline $\mathbf{B}_n$, with the single exception that $(n-1)$ always goes to the right-most column $(1\ n-1)$; if non-central, then $(1\ \alpha)$ goes to the column with central sequence that is adjacent from above to the column it would otherwise go to had it been central. It is convenient now to label the columns of each baseline $\mathbf{B}_n$ from left-to-right as $col_0, col_1, col_2, \ldots$, where the sequence terminates at a value characteristic of the baseline. The application of these rules effects **uniquely** the transformations between successive Creation Tables. This is presented in the following pictures for $n = 1 - 4$:

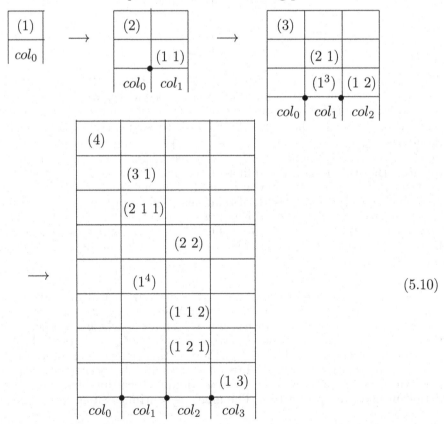

(5.10)

The point • in each of these diagrams marks the place where an MSS root in the set $\mathbb{MSS}$ of all MSS roots stands in the baseline $\mathbf{B}_n$.

The parametrization of the columns of the Creation Tables by col_i, $i = 0, 1, 2, \ldots$, has the property that the transformations between such tables can be given separately for general n for each column by the very simple transformation:

$$col_i \;\mapsto\; col_i, \quad i = 0, 1, 2, \ldots, \text{ for } (1\ \alpha) \text{ central;} \qquad (5.11)$$
$$col_i \;\mapsto\; col_{i+1}, \; i = 1, 2, \ldots, \text{ for } (1\ \alpha) \text{ noncentral.}$$

Here the central sequences are taken as known for general n, as shown earlier, so that each column in every table is covered exactly once in (5.11). It is also the case that the creation point • of all sequences in col_i is at the MSS root that corresponds to the left end of col_i. What could be simpler?

Summary. The parametrization of columns by $col_i, i = 0, 1, 2, \ldots$, in each Creation Table admits of a very simple enumeration of the sequences present in each such column of the Creation Table $\mathbb{T}_n$. Once these newly created sequences are known, even more properties of the unique deterministic-computer-generated inverse graphs $\mathbb{C}_n^*$ can be noted.

5.2 Properties of the Creation Table

The construction of the Creation Tables is one of the major accomplishments of this monograph. Since $\mathbb{T}_n$ contains the creation MSS roots of all sequences in the ordered set of 2^{n-1} positive sequences $\{(n), (n-1\ 1), \ldots, (1\ n-1)\}^{ord}$, hence, the conjugate sequences as well, it contains, in some sense, all properties of α-sequences and their conjugates. Several of the more important properties that can be read-off are illustrated below:

1. The set of all 2^n sequences in $\mathbb{T}_n \cup \overline{\mathbb{T}}_n$, including their order and creation points.

2. The set of all sequences created up to a specified value of ζ. This is the set of sequences contained in the merger (moving into one column) of all columns $col_i, i = 0, 1, \ldots, col_j$, where j is the column containing ζ. The merger of the remaining columns gives the set of all sequences yet to be created.

3. The set of all r-cycles. The determination of the sequences constituting an r-cycle is quite elusive, although they are unique and present in $\mathbb{T}_n$. Given $\mathbb{T}_n \cup \overline{\mathbb{T}}_n$, which of its sequences belong to the same cycle; at which MSS root were they created? It is the classification of columns of $\mathbb{T}_n$ in terms of the columns $col_i, i = 0, 1, 2, \ldots$: that provides a comprehensive answer. If the sequence $(1\ \alpha)$ is central for col_i, then col_i is the same (invariant) under the transformation $\mathbb{T}_{n-1} \mapsto \mathbb{T}_n$; otherwise, there is a shift $col_i \mapsto col_{i+1}$ upward to the adjacent column. Thus, an r-cycle either contains all cyclic permutations of the parts of the greatest sequence contained therein or the cyclic permutation of the least

sequence contained therein, where this rule applies to both positive and conjugate sequences. Thus, if $\mathbf{C}_n(\alpha_{\max}) = \alpha = (\alpha_0, \alpha_1, \ldots, \alpha_n) \in \mathbb{T}_n$ denotes the greatest sequence in a given set of r-cycles, then all cyclic permutations of the parts of $\mathbf{C}_n(\alpha_{\max})$ are in the same set of r-cycles, and this includes all sequences in the r-cycle. This gives

$$\mathbf{C}_n(\alpha_{\max}) = (\alpha_0, \alpha_1, \ldots, \alpha_n) \cup (\alpha_n, \alpha_0, \alpha_1, \ldots, \alpha_{n-1}) \cup \qquad (5.12)$$

$$(\alpha_{n-1}, \alpha_n, \alpha_0, \alpha_1, \ldots, \alpha_{n-2}) \cup \cdots (\alpha_1, \alpha_2, \ldots, \alpha_{n-1}, \alpha_0).$$

Thus, the notation $\mathbf{C}_n(\alpha_{\max})$ denotes the class of sequences equivalent to $\alpha_{\max} = (\alpha_0, \alpha_1, \ldots, \alpha_n)$ under cyclic permutations of its parts. Similarly, the notation $\mathbf{C}_n(\alpha_{\min}) = (\alpha'_0, \alpha'_1, \ldots, \alpha'_n)$ denotes the set of sequences

$$\mathbf{C}_n(\alpha_{\min}) = (\alpha'_n, \alpha'_{n-1}, \ldots, \alpha'_1, \alpha'_0) \cup (\alpha'_0, \alpha'_n, \ldots, \alpha'_2, \alpha'_1) \cup \cdots$$

$$(\alpha'_n, \alpha'_0, \alpha'_1, \ldots, \alpha'_{n-1}) \cup (\alpha'_0, \alpha'_1, \ldots, \alpha'_n). \qquad (5.13)$$

The procedure of effecting the method given above in (5.11)-(5.12) is the following: Cyclic permutations are applied to every sequence in the Creation Table $\mathbb{T}_n$, which then gives back the same table with a redistribution of its 2^n sequences. Then, if a given permuted sequence remains in the same col_i in both tables, it is assigned the same col_i in the redistribution; if it is transformed to a new col_j, $j \neq i$, it is assigned the new col_j. This procedure must then give all columns into which fall all the redistributed columns. The following results obtain:

$$\mathbf{C}_1((1)) \;=\; \{(1), \overline{(1)}\}\| \mapsto \{R, L\}; \qquad (5.14)$$

$$\mathbf{C}_2((2)) \;=\; \{(2), \overline{(1\ 1)} \mapsto \{RL, LR\};$$
$$\mathbf{C}_2(1^2) \;=\; \{(1\ 1)\} \mapsto \{RR\}; \qquad (5.15)$$
$$\mathbf{C}_2\overline{(1\ 1)} \;=\; \{\overline{1\ 1)}\} \mapsto \{LL\};$$

$$\mathbf{C}_3((3)) \;=\; \{(3), \overline{(12)}, \overline{(2\ 1)}\} \mapsto \{RLL, LRL, LLR\};$$
$$\mathbb{C}_3(2\ 1) \;=\; \{(2\ 1), (1\ 2), \overline{(1\ 1\ 1)}\} \mapsto \{RLR, RRL, LLR\},$$
$$\mathbf{C}_3((1^3)) \;=\; \{(1^3)\} \mapsto \{RRR\}; \qquad (5.16)$$
$$\mathbf{C}_3(\overline{(3)}) \;=\; \{\overline{(3)}\} \mapsto \{LLL\};$$

$$\mathbf{C}_4((4)) \;=\; \{(4), \overline{(1\ 3)}, \overline{(2\ 2)}, \overline{(3\ 1)}\} \mapsto \{RL^3, LRL^2, L^2RL, L^3R\};$$

$$\mathbf{C}_4((1^4)) \;=\; \{(1^4)\} \mapsto \{R^4\};$$

$$\mathbf{C}_4((1\ 2\ 1)) \;=\; \{(2\ 1\ 1), (1\ 1\ 2), (1\ 2\ 1), \overline{(1^4)}\}$$
$$\mapsto\; \{RLRR, RRRL, RRLR, LRRR\}; \qquad (5.17)$$

$$\mathbf{C}_4((1\ 3)) \;=\; \{(3\ 1), (1\ 3), \overline{(1\ 1\ 2)}, \overline{(2\ 1\ 1)}\}$$
$$\mapsto\; \{RLLR, RRLL, LRRL, LLRR\};$$

$$\mathbf{C}_4(\overline{(1\ 2\ 1)}) \;=\; \{(2\ 2), \overline{(1\ 2\ 1)}\} \mapsto \{RLRL, LRLR\};$$
$$\mathbf{C}_4(\overline{(4)}) \;=\; \{\overline{(4)}\} \mapsto \{L^4\}.$$

$$\mathbf{C}_5((5)) \;=\; \{(5), \overline{(1\ 4)}, \overline{(2\ 3)}, \overline{(3\ 2)}, \overline{(4\ 1)}\}$$
$$\mapsto \{RL^4,\ LRL^3,\ L^2RL^2,\ L^3RL,\ L^4R\};$$

$$\mathbf{C}_5((1^5)) \;=\; \{(1^5)\} \mapsto \{R^5\};$$

$$\mathbf{C}_5((1\ 2\ 2)) \;=\; \{(2\ 1\ 2), (2\ 2\ 1), (1\ 2\ 2), (2\ 1\ 2), \overline{(1\ 2\ 1\ 1)}, \overline{(1\ 1\ 2\ 1)}\}$$
$$\mapsto \{RLRRL,\ RLRLR,\ RRLRL,\ LRLRR, LRRLLR\};$$

$$\mathbf{C}_5((1\ 2\ 1\ 1)) \;=\; \{(2\ 1\ 1\ 1), (1\ 1\ 2\ 1), (1\ 1\ 1\ 2), (1\ 2\ 1\ 1), \overline{(1^5)}\}$$
$$\mapsto \{RLR^3,\ R^3LR,\ R^4L,\ R^2LRR, LR^4\};$$

$$\mathbf{C}_5((1\ 2\ 1\ 1)) \;=\; \{(2\ 1\ 1\ 1), (1\ 1\ 2\ 1), (1\ 1\ 1\ 2), (1\ 2\ 1\ 1), \overline{(1^5)}\}$$
$$\mapsto \{RLR^3,\ R^3LR,\ R^4L,\ R^2LRR, LR^4\}; \tag{5.18}$$

$$\mathbf{C}_5((1\ 3\ 1)) \;=\; \{(3\ 1\ 1), (1\ 1\ 3), (1\ 3\ 1), \overline{(1\ 1\ 1\ 2)}, \overline{(2\ 1\ 1\ 1)}\}$$
$$\mapsto \{RL^2R^2,\ R^3L^2,\ R^2L^2R,\ LR^3L, L^2R^3\};$$

$$\mathbf{C}_5((1\ 4)) \;=\; \{(4\ 1), (1\ 4), \overline{(1\ 1\ 3)}, \overline{(2\ 1\ 2)}, \overline{(3\ 1\ 1)}\}$$
$$\mapsto \{RL^3R,\ R^2L^3,\ LR^2L^2,\ L^2R^2L, L^3R^2\};$$

$$\mathbf{C}_5(\overline{(5)}) \;=\; \{\overline{(5)}\} \mapsto \{L^5\};$$

$$\mathbf{C}_5(\overline{(1\ 3\ 1)}) \;=\; \{(3\ 2), (2\ 3), \overline{(1\ 2\ 2)}, \overline{(1\ 3\ 1)}, \overline{(2\ 2\ 1)}\}$$
$$\mapsto \{RL^2RL,\ RLRL^2,\ LRLRL,\ LRL^2R, L^2RLR\}.$$

$$\mathbb{C}_6^*(\zeta_0) \;=\; \{(6)\};$$
$$\mathbb{C}_6^*(\zeta_1) \;=\; \{(5\ 1), (4\ 1^2), (3\ 1^3), (2\ 1^4), (1^6)\};$$
$$\mathbb{C}_6^*(\zeta_2) \;=\; \{(4\ 2), (3\ 1\ 2), (3\ 2\ 1), (2\ 1\ 2\ 1), (2\ 1\ 1\ 2), (2\ 2\ 2),$$
$$(1\ 1\ 2\ 2), (1\ 1\ 1\ 1\ 2), (1\ 1\ 1\ 2\ 1), (1\ 2\ 2\ 1)\};$$
$$\mathbb{C}_6^*(\zeta_3) \;=\; \{(2\ 2\ 1\ 1), (1\ 1\ 2\ 1\ 1), (1\ 2\ 1\ 1\ 1)\};$$
$$\mathbb{C}_6^*(\zeta_4) \;=\; \{(1\ 2\ 1\ 2)\}; \tag{5.19}$$
$$\mathbb{C}_6^*(\zeta_5) \;=\; \{(3\ 3), (2\ 1\ 3), (2\ 3\ 1), (1\ 1\ 3\ 1), (1\ 1\ 1\ 3), (1\ 2\ 3), (1\ 3\ 2)\};$$
$$\mathbb{C}_6^*(\zeta_6) \;=\; \{(1\ 3\ 1\ 1);$$
$$\mathbb{C}_6^*(\zeta_7) \;=\; \{2\ 4), (1\ 1\ 4), (1\ 4\ 1)\};$$
$$\mathbb{C}_6^*(\zeta_8) \;=\; (1\ 5).$$

In these last results for $n = 6$, the transformation to letters R and L is omitted to conserve space. All of these results, and others mentioned above, support the thesis that the creation tables may be taken as the basic elements in the development of the properties of α-sequences, with a simple, straightforward exposition of properties being contained in the col_i parametrization.

Example $n = 4$. Distribution of creation sequences:

(4)			
	(3 1)		
	(2 1 1)		
		(2 2)	
		(1 1 2)	
	(1^4)		
		(1 2 1)	
			(1 3)
$\longrightarrow$	$\longleftarrow$	$\longrightarrow$	$\longleftarrow$

$$0 \qquad \zeta((0)) \qquad \zeta((1)) \qquad \zeta((2)) \qquad 2$$

(5.20)

All labels of positive branches present in each central interval:

$$
\begin{aligned}
\mathbb{C}_6(\zeta_0, \zeta_1] &= \{(6)\}; \\
\mathbb{C}_6(\zeta_1, \zeta_2)] &= \{(6) \,|\, (1^6)\} - \mathbb{G}_6(\zeta_1, \zeta_2) = \mathbb{U}_6(\zeta_1, \zeta_2); \\
\mathbb{C}_6(\zeta_t, \zeta_{t+1}] &= \{(6) \,|\, c_n(t)\} - \mathbb{G}_6(\zeta_t, \zeta_{t+1}), \ t = 0, 1, \ldots, 8; \quad (5.21) \\
\mathbb{C}_6(\zeta_8, \infty] &= \{(6) \,|\, (1\ 5)\} \ (\text{full set}), \ \zeta_9 = \infty.
\end{aligned}
$$

The gap sequences in this relation are defined as follows:

$$
\begin{aligned}
\mathbb{G}_6(\zeta_1, \zeta_2) &= \sum_{s=1}^{3} \left\lfloor (5 - s\ 1^{s+1}) \,\middle|\, (4 - s\ 1^{s+2}) \right\rfloor = \left\lfloor (4\ 1^2) \,\middle|\, (3\ 1^3) \right\rfloor \\
&\quad + \left\lfloor (3\ 1^3) \,\middle|\, (2\ 1^4) \right\rfloor + \left\lfloor (2\ 1^4) \,\middle|\, (1^6) \right\rfloor + \left\lfloor (1^6) \,\middle|\, (1\ 5) \right\}; \\
\mathbb{G}_6(\zeta_2, \zeta_3) &= \left\lfloor {}^2(3\ 1^3)\,|\,(2\ 1^4) \right\rfloor^2 + \left\lfloor {}^3(2\ 1^4)\,|\,(1^6) \right\rfloor^3 \\
&\quad + \left\lfloor (1\ 2\ 2\ 1) \,\middle|\, (1\ 5) \right\}; \qquad\qquad\qquad\qquad (5.22a) \\
\mathbb{G}_6(\zeta_3, \zeta_4) &= \left\lfloor {}^2(3\ 1^3)\,|\,(2\ 1^4) \right\rfloor^2 + \left\lfloor {}^4(2\ 1^4) \,\middle|\, (1^6) \right\rfloor^4 + \left\lfloor (1\ 2\ 1\ 1\ 1) \,\middle|\, (1\ 5) \right\}; \\
\mathbb{G}_6(\zeta_4, \zeta_5) &= \left\lfloor {}^2(3\ 1^3)\,|\,(2\ 1^4) \right\rfloor^2 + \left\lfloor {}^4(2\ 1^4) \,\middle|\, (1^6) \right\rfloor^4 + \left\lfloor (1\ 2\ 1\ 2) \,\middle|\, (1\ 4) \right\}; \\
\mathbb{G}_6(\zeta_5, \zeta_6) &= \left\lfloor {}^5(2\ 1^4) \,\middle|\, (1^6) \right\rfloor^5 + \left\lfloor (1\ 3\ 2) \,\middle|\, (1\ 5) \right\}; \\
\mathbb{G}_6(\zeta_6, \zeta_7) &= \left\lfloor {}^5(2\ 1^4) \,\middle|\, (1^6) \right\rfloor^5 + \left\lfloor (1\ 3\ 1\ 1) \,\middle|\, (1\ 5) \right\}; \\
\mathbb{G}_6(\zeta_7, \zeta_8) &= \left\lfloor (1\ 4\ 1) \,\middle|\, (1\ 5) \right\} = \{(1\ 5)\}; \\
\mathbb{G}_6(\zeta_8, \infty) &= \left\lfloor (1\ 5) \,\middle|\, (1\ 5) \right\} = \text{empty sequence}.
\end{aligned}
$$

The gap sequences in these results are read-off the Table of Creation Sequences for $n = 6$ below by counting in from the end sequences:

$$\left\lfloor (4\ 1^2)\,|\,(3\ 1^3) \right\rfloor = \{(4\ 2),(3\ 1\ 2)\};$$

$$\left\lfloor (3\ 1^3)\,|\,(2\ 1^4) \right\rfloor = \{(3\ 2\ 1),(3\ 3),(2\ 1\ 3),(2\ 1\ 2\ 1)\};$$

$$\left\lfloor (2\ 1^4)\,|\,(1^6) \right\rfloor = \{(2\ 1\ 1\ 2),(2\ 2\ 2),(2\ 2\ 1\ 1),(2\ 3\ 1),(2\ 4),(1\ 1\ 4);$$

$$(1\ 1\ 3\ 1),(1\ 1\ 2\ 1\ 1),(1\ 1\ 2\ 2),(1\ 1\ 1\ 1\ 2)\}\};$$

$$\left\lfloor^2 (3\ 1^3)\,|\,(2\ 1^4) \right\rfloor^2 = \{(3\ 3),(2\ 1\ 3)\}; \tag{5.22b}$$

$$\left\lfloor^3 (2\ 1^4)\,|\,(1^6) \right\rfloor^3 = \{(2\ 2\ 1\ 1),(2\ 3\ 1),(2\ 4),(1\ 1\ 4),(1\ 1\ 3\ 1),(1\ 1\ 2\ 1\ 1)\};$$

$$\left\lfloor^4 (2\ 1^4)\,|\,(1^6) \right\rfloor^4 = \{(2\ 3\ 1),(2\ 4),(1\ 1\ 4),(1\ 1\ 3\ 1)\};$$

$$\left\lfloor^5 (2\ 1^4)\,|\,(1^6) \right\rfloor^5 = \{(2\ 4),(1\ 1\ 4)\}.$$

The subsequences of $\{(1^6)\,|\,(1\ 5)\}$ that enter into relations (5.22a) are those given in terms of central sequences as follows:

$$\left\lfloor (1\ 4\ 1)\,\middle|\,(1\ 5)\right\} = \{(1\ 5)\};$$

$$\left\lfloor (1\ 3\ 1\ 1)\,\middle|\,(1\ 5)\right\} = \left\{(1\ 4\ 1)\,\middle|\,(1\ 5)\right\};$$

$$\left\lfloor (1\ 3\ 2)\,\middle|\,(1\ 5)\right\} = \left\{(1\ 3\ 1\ 1)\,\middle|\,(1\ 5)\right\};$$

$$\left\lfloor (1\ 2\ 1\ 2)\,\middle|\,(1\ 5)\right\} = \left\{(1\ 3\ 2)\,\middle|\,(1\ 5)\right\}; \tag{5.23}$$

$$\left\lfloor (1\ 2\ 1\ 1\ 1)\,\middle|\,(1\ 5)\right\} = \left\{(1\ 2\ 1\ 2)\,\middle|\,(1\ 5)\right\};$$

$$\left\lfloor (1\ 2\ 2\ 1)\,\middle|\,(1\ 5)\right\} = \left\{(1\ 2\ 1\ 1\ 1)\,\middle|\,(1\ 5)\right\};$$

$$\left\lfloor (1^5)\,\middle|\,(1\ 5)\right\} = \left\{(1\ 1\ 1\ 2\ 1),(1\ 1\ 1\ 3),(1\ 2\ 3)\right\}$$

$$+ \left\{(1\ 2\ 2\ 1)\,\middle|\,(1\ 5)\right\}.$$

Cycle class creation:

There are fourteen cycle classes: nine 6-cycles, two 3-cycles, one 2-cycle, and two 1-cycles. Because of space considerations, these are listed here by giving just the central sequence representative of each cycle class; the full set is then completed by effecting the cyclic permutations of the representative and reading back the corresponding sequence, as given by:

6-cycle: $\mathbf{C}_6((6)) = \{(6), \ldots\} \mapsto \{RL^5, \ldots\};$

1-cycle: $\mathbf{C}_6((1^6)) = \{(1^6)\} \mapsto \{R^6\};$

6-cycle: $\mathbf{C}_6((1\ 2\ 2\ 1)) = \{(1\ 2\ 2\ 1), \ldots\} \mapsto \{RRLRLR, \ldots\};$

6-cycle: $\mathbf{C}_6((1\ 2\ 1\ 1\ 1)) = \{(1\ 2\ 1\ 1\ 1), \ldots\} \mapsto \{RRLRRR, \ldots\};$

3-cycle: $\mathbf{C}_6((1\ 2\ 1\ 2)) = \{(1\ 2\ 1\ 2), \ldots\} \mapsto \{RRLRRL, \ldots\};$

6-cycle: $\mathbf{C}_6((1\ 3\ 2)) = \{(1\ 3\ 2), \ldots\} \mapsto \{RRLLRL, \ldots\};$

6-cycle: $\mathbf{C}_6((1\ 3\ 1\ 1)) = \{(1\ 3\ 1\ 1), \ldots\} \mapsto \{RRLLRR, \ldots\};$

6-cycle: $\mathbf{C}_6((1\ 4\ 1)) = \{(1\ 4\ 1), \ldots\} \mapsto \{RRLLLR, \ldots\};$ (5.24)

6-cycle: $\mathbf{C}_6((1\ 5)) = \{(1\ 5), \ldots\} \mapsto \{RRLLLL, \ldots\};$

1-cycle: $\mathbf{C}_6(\overline{(6)}) = \{\overline{((6))}\} \mapsto \{L^6\};$

2-cycle: $\mathbf{C}_6(\overline{(1\ 2\ 2\ 1)}) = \{\overline{(1\ 2\ 2\ 1)}\} \mapsto \{LRLRLR, \ldots\};$

3-cycle: $\mathbf{C}_6(\overline{(1\ 3\ 2)}) = \{\overline{(1\ 3\ 2)}, \ldots\} \mapsto \{LRLLRL, \ldots\},$

6-cycle: $\mathbf{C}_6(\overline{(1\ 3\ 1\ 1)}) = \{\overline{(1\ 3\ 1\ 1)}\} \mapsto \{LRLLRR, \ldots\};$

6-cycle: $\mathbf{C}_6(\overline{(1\ 4\ 1)}) = \{\overline{(1\ 4\ 1)}, \ldots\} \mapsto \{LRLLLR, \ldots\}.$

The creation points of the members of these cycle classes are the MSS roots that stand at the left-end of the column of the central sequence in which a label in a given cycle class occurs in the Table of Creation Sequences for $n = 6$ below. This long listing is omitted for $n = 6$, since it is fully illustrated in the previous examples for $n = 2, \ldots, 5$.

A well-known property of r-cycles is expressed by relations (5.9) in that the cyclic permutations of each element of an r-cycle gives back the same r-cycle.

It is also useful to note the number of cycle classes in set of all cycles for a given n. This is illustrated for $n = 5$ by the following sets:

Cycle class creation:

There are eight cycle classes: six 5-cycles and two 1-cycles:

$$\mathbf{C}_5((5)) \ = \ \{(5), \overline{(1\ 4)}, \overline{(2\ 3)}, \overline{(3\ 2)}, \overline{(4\ 1)}\}$$
$$\mapsto \ \{RL^4, LRL^3, L^2RL^2, L^3RL, L^4R\};$$
$$\mathbf{C}_5((1^5)) \ = \ \{(1^5)\} \mapsto \{R^5\};$$
$$\mathbf{C}_5((1\ 2\ 2)) \ = \ \{(2\ 1\ 2), (2\ 2\ 1), (1\ 2\ 2), (2\ 1\ 2), \overline{(1\ 2\ 1\ 1)}, \overline{(1\ 1\ 2\ 1)}\}$$
$$\mapsto \ \{RLRRL, RLRLR, RRLRL, LRLRR, LRRLLR\};$$
$$\mathbf{C}_5((1\ 2\ 1\ 1)) \ = \ \{(2\ 1\ 1\ 1), (1\ 1\ 2\ 1), (1\ 1\ 1\ 2), (1\ 2\ 1\ 1), \overline{(1^5)}\}$$
$$\mapsto \ \{RLR^3, R^3LR, R^4L, R^2LRR, LR^4\};$$

$$\mathbf{C}_5((1\ 3\ 1)) = \{(3\ 1\ 1), (1\ 1\ 3), (1\ 3\ 1), \overline{(1\ 1\ 1\ 2)}, \overline{(2\ 1\ 1\ 1)}\} \qquad (5.25)$$
$$\mapsto \{RL^2R^2,\ R^3L^2,\ R^2L^2R,\ LR^3L, L^2R^3\};$$
$$\mathbf{C}_5((1\ 4)) = \{(4\ 1), (1\ 4), \overline{(1\ 1\ 3)}, \overline{(2\ 1\ 2)}, \overline{(3\ 1\ 1)}\}$$
$$\mapsto \{RL^3R,\ R^2L^3,\ LR^2L^2,\ L^2R^2L, L^3R^2\};$$
$$\mathbf{C}_5(\overline{(5)}) = \{\overline{(5)}\} \mapsto \{L^5\};$$
$$\mathbf{C}_5(\overline{(1\ 3\ 1)}) = \{(3\ 2), (2\ 3), \overline{(1\ 2\ 2)}, \overline{(1\ 3\ 1)}, \overline{(2\ 2\ 1)}\}$$
$$\mapsto \{RL^2RL,\ RLRL^2,\ LRLRL,\ LRL^2R, L^2RLR\}.$$

Further Examples. It is universal that $(n), (1^n), (1\ n-1)$ are always central sequences in $\mathbb{T}_n$, obtained, respectively by application of the $+1$-rule to $(n-1)$, and the application of the $(1\ \alpha)$-rule to $(1^{n-1}, (n-1))$. This is taken into account in the following enumeration of sequences in $\mathbb{T}_n$:

(1). Central sequences for $n = 4$ as found from $\alpha \in \mathbb{T}_3$. The only sequence in $\mathbb{T}_3$ that needs to be considered is the sequence $(2\ 1)$, which maps to the two sequences $(3\ 1), (1\ 2\ 1) \in \mathbb{T}_4$, and only $(1\ 2\ 1)$ qualifies as a possible central sequence. The solution of $\Lambda_4(\gamma) = (2\ 1)$ is $\gamma = (1)$, which gives the MSS root $\zeta((1))$ as the creation value of the corresponding central branch function $\Psi_\zeta((1\ 2\ 1); x)$. Thus, the placement of all central sequences in baseline $\mathbb{B}_4$ is uniquely obtained in this manner from the given baseline $\mathbb{B}_3$.

(2). Central sequences for $n = 5$ as found from $\alpha \in \mathbb{T}_4$. The only sequences in $\mathbb{T}_4$ that needs to be considered is the set of sequences

$$\{(3\ 1),\ (2\ 1\ 1),\ (2\ 2),\ (1\ 1\ 2),\ (1\ 2\ 1)\}. \qquad (5.26)$$

It is only the $(1\ \alpha)$-rule applied to these sequences that gives a set of potential central sequences in $\mathbb{T}_5$:

$$\{(1\ 3\ 1),\ (1\ 2\ 1\ 1),\ (1\ 2\ 2),\ (1\ 1\ 1\ 2),\ (1\ 1\ 2\ 1)\}. \qquad (5.27)$$

Of these candidates for central sequences only the first three have a lexical solution $\gamma \in L_1, L_2, L_4, L_5$ such that

$$\Lambda_5(\gamma) = \alpha,\ \alpha \in \{(3\ 1),\ (2\ 1\ 1),\ (2\ 2)\}. \qquad (5.28)$$

The three respective lexical sequences are: $\gamma = (2), (2\ 1), (1)$, which give the MSS roots $\zeta((2)), \zeta((2\ 1)), \zeta((1))$ as the creation values of the corresponding central branch functions $\Psi_\zeta((1\ 3\ 1); x), \Psi_\zeta((1\ 2\ 1\ 1); x), \Psi_\zeta((1\ 2\ 2); x)$. Thus, the placement of all central sequences in baseline $\mathbb{B}_5$ is uniquely obtained in this manner from the given baseline $\mathbb{B}_4$.

Examples such as those given above serve to illustrate the many guises under which the unique deterministic-computer-generated inverse sequences $\mathbb{C}_n^*$ present themselves.

5.2.1 The Table of Creation Sequences for Prime n

The computational structural aspects of the set $\mathbb{C}_n^*$ is concluded in this last section of Chapter 5 by discussion of (i) the unexpected role of prime numbers; (2) a quite understandable formating of the set $\mathbb{C}^*$ showing all sequences $\mathbb{C}_n^*$ for $n = 1, 2, \ldots, 6$.

(i). The number of lexical sequences in the set L_n of lexical sequences of degree $n - 1$ defined by (see Sect. 1.2.5)

$$L_n = \{\alpha = (\alpha_0, \alpha_1, \ldots, \alpha_k) \mid \alpha_0 + \alpha_1 + \cdots + \alpha_k = n - 1; \alpha \text{ lexical}\} \quad (5.29)$$

is given by:

$$|L_n| = \frac{2^{n-1} - 1}{n}, \quad (5.30)$$

where Fermat's theorem assures that the n divides $2^{n-1} - 1$ for each prime number n $(n > 2)$.

The central curves for n prime are given by:

$$C_\zeta^n\left((n)\mid \overline{(n)}\right), \ \zeta \in (0, \zeta((0))];$$

$$C_\zeta^n\left((1^n)\mid \overline{(1^n)}\right), \ \zeta \in (\zeta((0)), \zeta_2];$$

$$C_\zeta^n\left((1 \ \alpha^{(t)})\mid \overline{(1 \ \alpha^{(t)})}\right), \ \alpha^{(t)} \in L_n; \quad (5.31)$$

$$\Lambda_n(\gamma^{(t)}) = \alpha^{(t)}, \ \zeta \in (\zeta_t, \zeta_{t+1}];$$

$$\gamma^{(t)} \in \{L_2, L_3, \ldots, L_{n-1}\}, \ t = 2, 3, \ldots, |L_n| + 1.$$

The MSS root $\zeta_t = \zeta(\gamma^{(t)})$ at which the central curve $C_\zeta^n\left(1 \ \alpha^{(t)} \mid \overline{(1 \ \alpha^{(t)})}\right)$ for the interval $(\zeta_t, \zeta_{t+1}]$ is created is determined by considering all sequences $\Lambda_n(\gamma^{(t)})$, $\gamma^{(t)} \in \{L_2, L_3, \ldots, L_{n-1}\}$, and then selecting the subset that gives the lexical sequences $\alpha^{(t)} \in L_n$. The following results emerge from this:

$$\text{number of sequences above } 1^n = 2\frac{2^{n-1} - 1}{3};$$

$$(5.32)$$

$$\text{number of sequences below } 1^n = \frac{2^{n-1} - 1}{3}.$$

It is also known for n prime that all cycles classes are n-cycles, except for the two 1-cycles corresponding to $(1^n) \mapsto R^n$ and $(-n) \mapsto L^n$. Hence, it must be the case that $2^n - 2 = nX_n$, where X_n denotes the number of n-cycles of length n for prime n; hence, the number X_n is given by

$$X_n = 2|L_n|. \quad (5.33)$$

(ii). The ζ-values for the creation of new labels of the positive branches of the inverse graph G_ζ^n can be pictured in stacked arrays of labels (sequences) in the displays below: Despite the space it requires to set forth these diagrams clearly, it is worth it because it unveils the general structure underlying the relationship between central sequences and the creation of new sequences. The examples are given for $n = 1, 2, \ldots, 6$. The diagrams all have the following structure: Each display has 2^{n-1} rows and a number of columns equal to the number $|\mathbb{C}_n|$ of central sequences. Each row contains one label with the greatest label (n) in the top row and the least label $(1\ n-1)$ in the bottom row; hence, each label in the set $\{(n) \,|\, (1\ n-1)\}$ appears exactly once in some column. It is the columns that contain the more relevant structural information. Each column contains the set of all new labels of the branches of G_{ζ}^n created synchronously at ζ_t with the central curve for the interval $(\zeta_t, \zeta_{t+1}], t = 0, 1, \ldots, |\mathbb{C}_n| = q_n + 1$. The order of the stacked elements in each column is always from greatest-to-least as read from top-to-bottom, and the least element is always that of the positive branch of the central sequence for that interval.

The information displayed in each of the stacked displays in $\mathbb{T}_n$ is exactly the distribution of the new labels into sets associated with each interval $\mathbb{C}_n(\zeta_t, \zeta_{t+1}]$. The so-called gap sequences can also be read off directly from the given stacked displays, as next described.

Examples. Several examples of the information contained in the inverse graph G_{ζ}^n that can be read-off these creation tables are:

1. The full ordered membership of the set $\{(n) \,|\, (1\ n-1)\}$.

2. The set of labels of the positive central branches created at the MSS root ζ_t of each interval $(\zeta_t, \zeta_{t+1}], t = 0, 1, \ldots, q_n$, including the central branch. These are the labels in the column $0, 1, \ldots, t, \ldots, q_n$.

3. The set of ordered labels from top-to-bottom of **all branches** in the inverse graph G_{ζ}^n for all $\zeta \in (\zeta_t, \zeta_{t+1}]$. This is the set of labels obtained by **merging together into one column** all of the entries in col_i, $i = 0, 1, \ldots, t$.

4. The gap labels yet to be created in the further ζ-evolution. This is the set of labels in the full set $\{(n) \,|\, (1\ n-1)\}$, but still missing from the merger described in Item 3:

$$\mathbb{G}_n(\zeta_t, \zeta_{t+1}) = \text{sum of all sequences in columns}$$
$$t + 1 \text{ to } q_n, t = 1, 2, \ldots, q_n - 1. \quad (5.34)$$

Further Examples. The examples below also show the direction of motion of each central branch:

$n = 1$. Distribution of creation sequences:

$$(5.35)$$

$n = 2$. Distribution of creation sequences:

(2)	
	(1 1)
$\longrightarrow$	$\longleftarrow$

0           $\zeta((0))$ 2

$$(5.36)$$

$n = 3$. Distribution of creation sequences:

(3)		
	(2 1)	
	(1 1 1)	
		(1 2)
$\longrightarrow$	$\longrightarrow$	$\longleftarrow$

0           $\zeta((0))$ $\zeta((1))$ 2

$$(5.37)$$

$n = 4$. Distribution of creation sequences:

(4)			
	(3 1)		
	(2 1 1)		
		(2 2)	
		(1 1 2)	
	(1^4)		
		(1 2 1)	
			(1 3)
$\longrightarrow$	$\longleftarrow$	$\longrightarrow$	$\longleftarrow$

$$\begin{array}{ccccc} 0 & \zeta((0)) & \zeta((1)) & \zeta((2)) & 2 \end{array}$$

$$(5.38)$$

$n = 5$. Distribution of creation sequences:

(5)					
	(4 1)				
	$(3\ 1^2)$				
		(3 2)			
		(2 1 2)			
	$(2\ 1^3)$				
		(2 2 1)			
				(2 3)	
				(1 1 3)	
		(1 1 2 1)			
	(1^5)				
		(1 1 1 2)			
		(1 2 2)			
			(1 2 1 1)		
				(1 3 1)	
					(1 4)
$\longrightarrow$	$\longrightarrow$	$\longrightarrow$	$\longleftarrow$	$\longrightarrow$	$\longleftarrow$

$$\begin{array}{ccccccc} 0 & \zeta((0)) & \zeta((1)) & \zeta((2\ 1)) & \zeta((2)) & \zeta((3)) & 2 \end{array}$$

$$(5.39)$$

Example $n = 6$. Distribution of creation sequences:

| | ζ((0)) | ζ((1)) | ζ((2 1)) | ζ((2 1 1)) | ζ((2)) | ζ((3 1)) | ζ((3)) | ζ((4)) |
→	←	←	→	←	→	←	→	←
(6)								
	(5 1)							
	$(4\ 1^2)$							
		(4 2)						
		(3 1 2)						
	$(3\ 1^3)$							
		(3 2 1)						
					(3 3)			
					(2 1 3)			
		(2 1 2 1)						
	$(2\ 1^4)$							
		(2 1 1 2)						
		(2 2 2)						
			(2 2 1 1)					
					(2 3 1)			
							(2 4)	
							(1 1 4)	
					(1 1 3 1)			
				(1 1 2 1 1)				
		(1 1 2 2)						
		(1 1 1 1 2)						
	(1^6)							
				(1 1 1 2 1)				
					(1 1 1 3)			
					(1 2 3)			
		(1 2 2 1)						
				(1 2 1 1 1)				
				(1 2 1 2)				
					(1 3 2)			
						(1 3 1 1)		
							(1 4 1)	
								(1 5)

0 (left endpoint) ... 2 (right endpoint)

$$(5.40)$$

The creation tables in (5.35)-(5.40) above show in vivid detail how these tables are generated recursively, one table at a time, each from the preceding table. It is useful to repeat these result in terms of the col_i nomenclature so that it is also verified directly from (5.9) for $n = 1 - 4$:

Select a given Creation Table $\mathbb{T}_n$. Then, this table is constructed from $\mathbb{T}_{n-1}$ by effecting all cyclic permutations of the sequences in each col_i. If the permuted sequence occurs in the same col_i as the original sequence, then it remains in the same col_i in Creation Table $\mathbb{T}_n$; if the permuted sequence occurs in a col_j, $j \neq i$, then it is transferred into col_j in Creation Table $\mathbb{T}_n$.

Select a given Creation Table $\mathbb{T}_n$. Then, the sequences that fall into the same fixed set as an arbitrarily selected sequence from $\mathbb{T}_n$ are obtained by cyclic permutations of that sequence. The sequences in each such set are uniquely labeled either by the greatest sequence or the least sequence contained therein. Thus, the fixed points belong to one or the other of the equivalence classes of sequences denoted by $\mathbf{C}_n(\alpha_{\max})$, or $\mathbf{C}_n(\alpha_{\min})$, or both should $\alpha_{\max} = \alpha_{\min}$ (a 1-cycle).

Another very important feature of the inverse graph is shown by the vectors placed in each interval of the baseline. These vectors indicate the **direction of motion of the central curve for that interval.** Thus, starting from the left-most boundary at $\zeta = 0$, this motion with increasing ζ is described as follows: The motion is along the horizontal central line $y = 1$ of the inverse graph, back and forth through the fixed central point $(1, 1)$ of the graph, with a variable amplitude that depends on the extremal points of the right-moving or left-moving central curve as specified by the arrow for that MSS interval. This oscillatory behavior holds for all $\zeta \in (0, 2]$, but the last left-moving central curve with its left-moving motion continues leftward for all $\zeta \in [2, \infty)$; that is, the central curve for each $n \geq 2$ is ejected to the left of the finite interval $[0, 2]$, where it continues leftward forever, that is, for all $\zeta \geq 2$. The details of this motion are given in the computer graphs of the MSS polynomials $q_n(\zeta)$ defined in Sect. 1.2.4, as presented in the graphs labeled P2(ZETA)-P10(ZETA), with $Pn(\text{ZETA}) = q_n(\zeta)$, where ZETA $= \zeta$ denote the same parameter. While no new fixed points are created for $\zeta \geq 2$, the features of the inverse graph continue to evolve in perhaps unexpected ways, as next discussed.

The phenomenon in question is also shown quite vividly in the inverse graphs labeled by $P8, \zeta = 2.00000, \zeta = 2.00001, \zeta = 2.00002$. It may be described for general n. The notation $\{(k + 1 \quad 1^{n-k-1}) \mid (k \quad 1^{n-k})\}\rceil$ is well-suited for this description, where it is recalled that this notation with $\rceil$ to the right of the ordered set of labels $\{(k + 1 \quad 1^{n-k-1}) \mid (k \quad 1^{n-k})\}$ designates that the right-most element $(k \quad 1^{n-k})$ is removed. The notation $\{\alpha \mid \alpha'\}, \alpha \geq \alpha'$, itself denotes the ordered set of all elements in the set $\mathbb{A}_n$ of 2^{n-1} positive elements that fall between α and $\alpha,'$ including these two end sequences.

The phenomenon referred to above in the inverse graphs $P8$ are described for general n in terms of the notations above, where now the following abbreviated notation is also introduced: $\mathcal{F}_k^n = \{(k+1 \quad 1^{n-k-1}) \mid (k \quad 1^{n-k})\}\rceil$, $k = 1, 2, \ldots, n-1$. Thus, for general n, there are $n-1$ such families of sequences given schematically by the following picture:

$$\mathcal{F}_{n-1}^n \quad \underline{\hspace{6cm}}$$

$$\mathcal{F}_{n-2}^n \quad \underline{\hspace{6cm}}$$

$$\mathcal{F}_1^n \quad \underline{\hspace{5cm}} \tag{5.41}$$

$$\vdots \qquad \vdots$$

$$\mathcal{F}_0^n \quad \underline{\hspace{5cm}}$$

The sequences $\mathcal{F}_0^n = \{(1^n), \ldots, (1 \quad n - 1)\}^{ord}$ must be adjoined to capture the maximal sequence (1^n). The individual branch sequences $\mathcal{F}_k^n$ are not resolved in the computer-generated graphs referred to above and certainly do not coincide exactly for the indicated finite values of ζ. Moreover, the gap between families does appear to be present. These spacings all evolve in ζ. Notice also that the greatest sequence in the family $\mathcal{F}_k^n$ is $(k{+}1\ \ 1^{n-k+1})$. The collection of greatest sequences is $\{(n),(n{-}1\ 1),(n{-}2\ 1^2),\ldots,(2\ 1^{n-1}),(1^n)\}$ is just the universal set discussed in Sect. 1.2. This regularity of structure for general n reinforces the interpretation of the unresolved computer-generated graphs.

It is useful to give the sequences in the picture (5.42) in the case $n = 8$:

$$
\begin{aligned}
\mathcal{F}_7^8 &= \{(8)\}, \\
\mathcal{F}_6^8 &= \{(7\ 1)\}, \\
\mathcal{F}_5^8 &= \{(6\ 1^2),(6\ 2),(5\ 1\ 2)\}^{ord}, \\
\mathcal{F}_4^8 &= \{(5\ 1^3),(5\ 2\ 1),(5\ 3),(4\ 1\ 3),(4\ 1\ 2\ 1)\}^{ord}, \qquad (5.42)\\
\mathcal{F}_3^8 &= \{(4\ 1^4),\ldots,(3\ 1\ 1\ 1\ 2)\}^{ord}, \\
\mathcal{F}_2^8 &= \{(3\ 1^5),\ldots,(2\ 1\ 1\ 1\ 2\ 1)\}^{ord}, \\
\mathcal{F}_1^8 &= \{(2\ 1^6),\ldots,(1^6\ 2)\}^{ord}, \\
\mathcal{F}_0^8 &= \{(1^8),\ldots,(1\ 7)\}^{ord}.
\end{aligned}
$$

The above structure for ζ outside the interval $[0,2]$ is only the beginning of, perhaps, unexpected behavior of the motions of the branches of the inverse graph. The following quite surprising events seems to occur. The computer-generated graphs $P\,3$ for negative ζ given by $\zeta = -1.10000$, -1.00000, -0.90000, -0.80000, -0.70000, -0.60000, -0.50000, -0.40000, -0.30000, -0.20000, -0.10000, 0.00000 suggest that for ζ very large and negative the positive branches of the inverse graph are at a finite value of ζ on the positive side of the inverse graph and at a symmetrically placed horizontal line on the negative (conjugate) side. Then, as ζ increases to the right, these lines move into the band-like sets of sequences $\mathcal{F}_k^n$, and these sets of sequences continue to evolve continuously to greater y-values on the positive side with the number of branches decreasing to exactly two at $\zeta = 0$, where it contracts to the starting position of the entire inverse graph for all $\zeta > 0$. This "reversal process" for negative values of ζ is quite speculative and requires further computation. The computer-generated graphs for $P\,8$, where n is even, seem to follow a quite different path in the negative ζ-domain in reaching the starting point at $\zeta = 0$, where the replication of the entire inverse graph begins. In this case, at large negative values of x, the branches of the inverse graph appear to be distributed from very large positive y-values downward to very large negative y-values, symmetrically placed about the $y = 1$ centerline. As x increases to the right, the curves constituting the inverse graph move continuously and symmetrically toward the center line $y = 1$, matching up exactly with the $x = 0$ starting line of the entire inverse graph for $x > 0$. These motions of the inverse graph for negative ζ-values are highly speculative, since there are not enough computer-generated graphs to be convincing. At the time the computer-generated graphs $\mathbb{C}_n^*$ were done the extension to ζ-values outside the interval $[0,2]$ was just a curiosity; it was not carried out. It should have been, since now it seems likely that the set of universal inverse graphs plays a major structural role.

There are many inverse graphs displayed in the long Chapter 7. The values of ζ given on each diagram of an inverse graph were selected for the purpose of showing various features of the inverse graph, which should be visualized as constituting discrete snapshots of a continuously evolving graph. It is quite impossible to direct attention to specific details in each graph. Guidelines are to look for changes in the rightward and leftward motions as ζ increases, as well as the number of graphs that appear, which is always increasing with order preservation and with no crossings. Many less-than-visible features have also been identified at various places throughout this monograph. Careful examination and extrapolation from the presented inverse graphs often allow their verification. Also, the selection given at the various ζ-values is intended to exhibit computationally the richness of structure that qualifies the subject of this monograph as a complex system.

Applications of the viewpoint of chaos theory presented in this monograph were made in Refs. [40–44] prior to the discovery that it has a recursion-based structure uniquely given by the deterministic-computer-generated inverse graph $\mathbb{C}_n^*$. It may well be that this result implies further properties of the subjects of Refs. [40–44], but this has been put aside in favor of giving quicker dissemination of the subject of Chapter 4. Other viewpoints of chaos theory are references principally for the purpose of illustrating the diversity of viewpoints of what is called chaos theory, thereby showing its importance as a new branch of physics that often gives unexpected insights into the behavior of physical systems.

Lastly, the important issues of computability and reproducibility of the inverse graph raised by Lorenz [27] are not issues in this monograph. The methods of double precision implemented by Bivens and Stein in Ref. [20] are such that all inverse graphs herein presented are reproducible; this is because the starting point for computing the inverse graph incrementally can be taken as the **exact** point $x = 1$ or $x = 3/2$. One, of course, would never choose one of increments as a starting point, as advised already by Lorenz [27]. The fact that also all inverse graphs meet $x = 0$ and $x = 2$ at exactly $\zeta = 2$ further confirms the reproducibility. Still another feature is the **invariance** of fixed points between the inverse graphs G_ζ^n and the ordinary graphs H_ζ^n. This latter property can be expressed by the composition of functions given by

$$G_\zeta^n \circ H_\zeta^n = H_\zeta^n \circ G_\zeta^n = \mathbb{I}_\zeta, \tag{5.43}$$

where $\mathbb{I}_\zeta$ is the identity function for function composition, and relation (5.43) holds independently of x.

Chapter 6

Graphical Presentation of MSS Roots

This Chapter presents a computer-generated graph of the curve whose zeros give the MSS roots (see relations (1.66)) for $n = 2, 3, \ldots, 10$. Thus, the graph presented below in terms of the notation $P\,n(\text{ZETA})$ is the polynomial (1.66) given by $p_6(\zeta) = P\,n(\text{ZETA})$. Each graph shows the intersection of this curve with the central axis at $y = 1$, this intersection point being the MSS root in question from which the approximate value of the root can be read-off. These graphs show vividly the number of MSS roots (all positive roots) for $n = 1, 2, \ldots, 7$, as well as the dramatic pile-up of roots as ζ approaches value 2, where they are unresolved by the computer print-out for $n = 8, 9, 10$. (Thanks are given to Alpha Graphics of Santa Fe for providing page numbers and headings to the graphs that comprise the raw-data computer print-outs used in this monograph.)

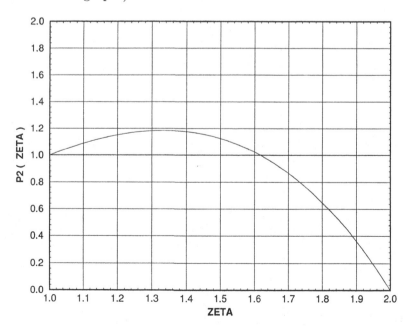

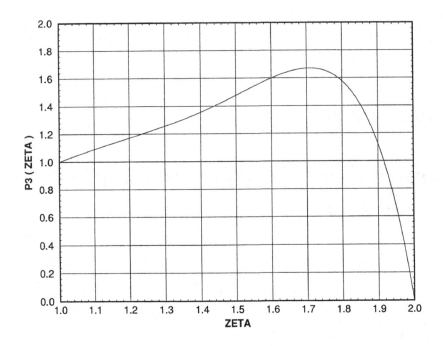

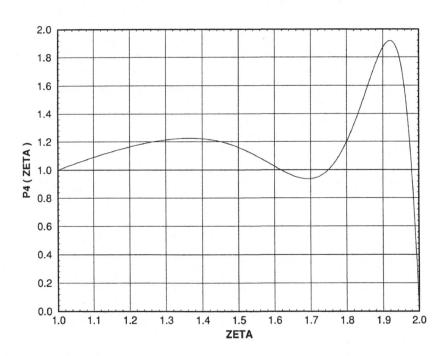

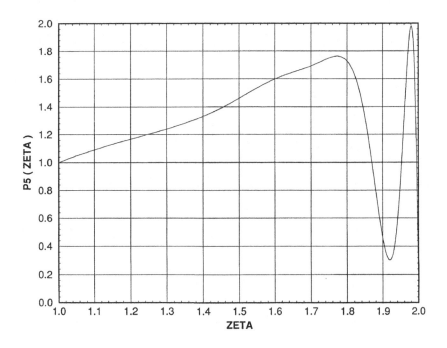

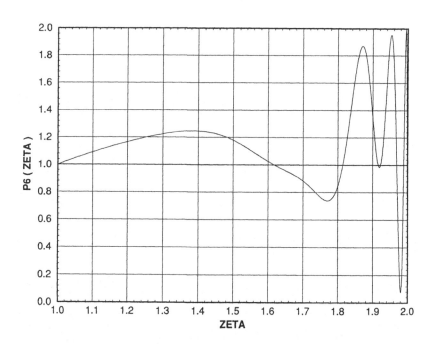

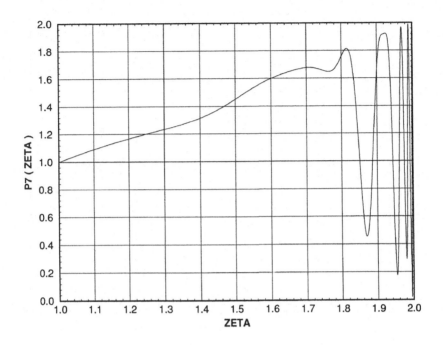

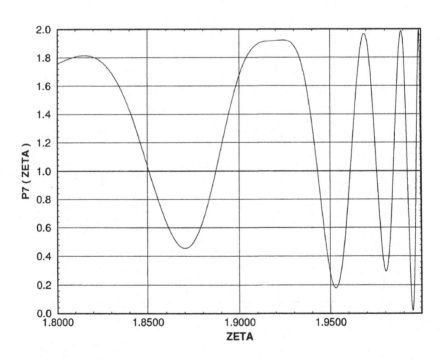

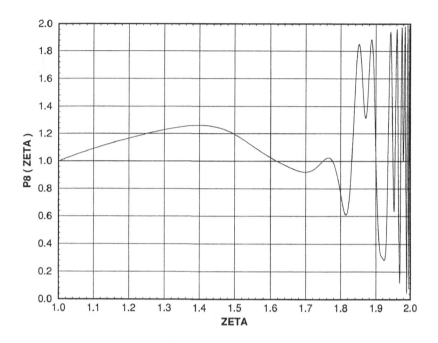

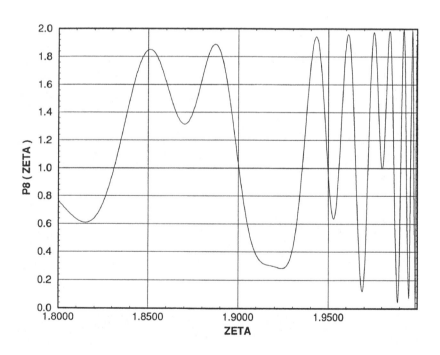

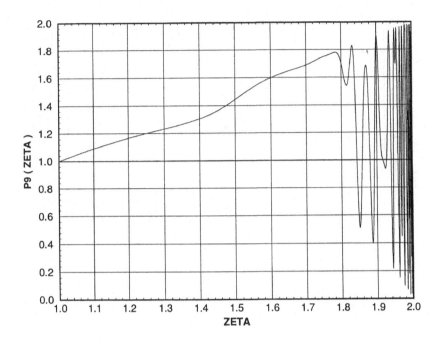

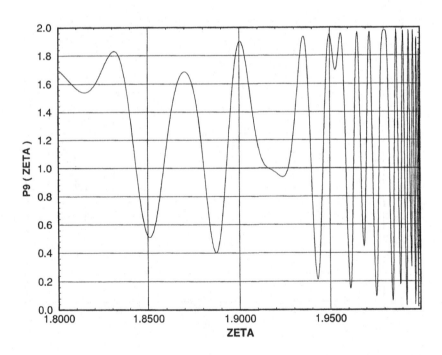

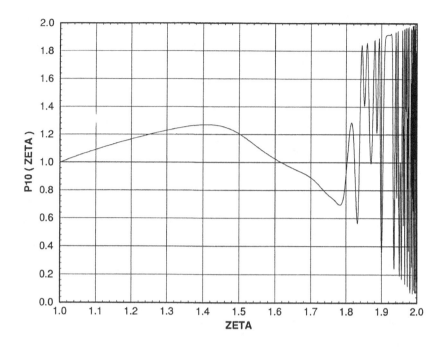

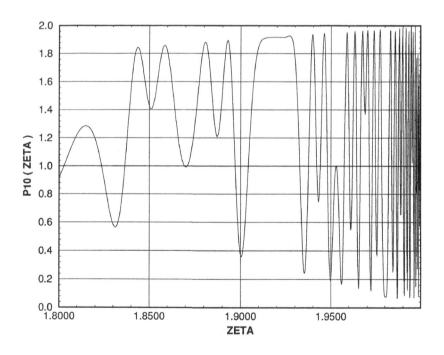

Chapter 7

Graphical Presentation of Inverse Graphs

This Chapter contains the following seven graphs denoted by $P2$, $P3$, $P4$, $P5$, $P6$, $P8$, $P16$, which is the label just below the baseline at the bottom of each graph. The number of inverse graphs with this label that are presented is given by the integer in parentheses as follows: $P2(6), P3(12), P4(12), P5(11), P6(43), P8(34), P16(36)$, for a total of 154 inverse graphs. This may seem excessive, but each graph is intended to show features of the inverse graph that are important for understanding their ζ-evolution and to establish that:

The collection of all inverse graphs is a complex adaptive system in which the operation of function composition is the basic mathematical rule.

Principal features to look for are the creation of fixed points and new branches, the left-right motion of the branches and their extremal points, especially, the creation of central sequences at the various MSS roots and the synchronous creation of noncentral branches. In the "mind's-eye," it is useful to visualize the various snapshots as a smooth-running picture of the creation of the abstract space for each $n = 1, 2, \ldots$. In retrospect, probably a more informative collection would be presented, but at the time these graphs were generated the computational techniques available were intertwined with theory to find a way to the correct picture. The readers will, no doubt, develop their own methods of looking for and correctly extrapolating between snapshots.

A precautionary note is that very careful attention must be paid to the (x, y)-coordinate domains of definition that occur in various graphs in order to focus-in on local features, which are, in fact, always found to verify the theory; that is:

Chaos Theory is proved to be a Complex Adaptive System, a result that is verified by Computer Proof.

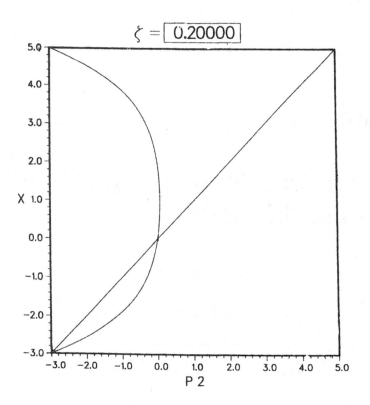

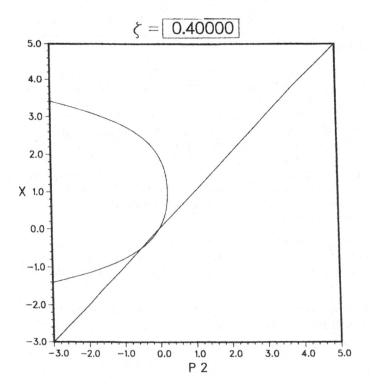

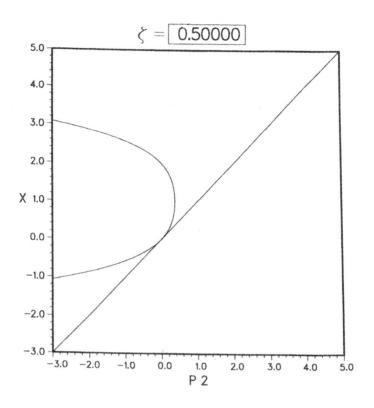

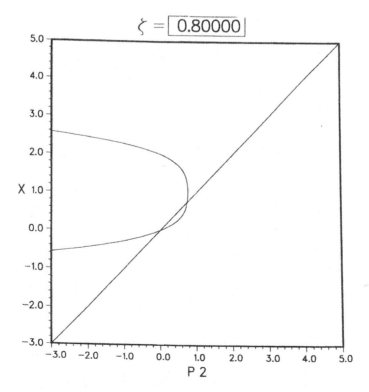

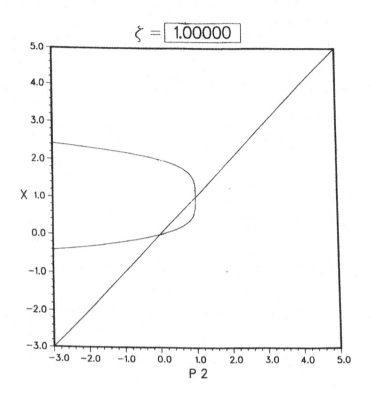

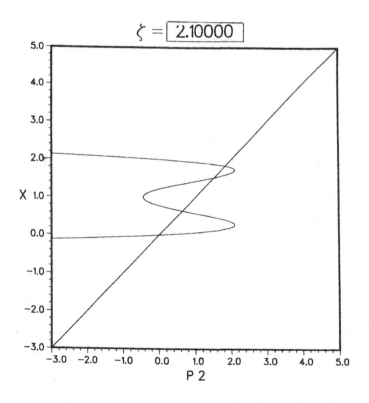

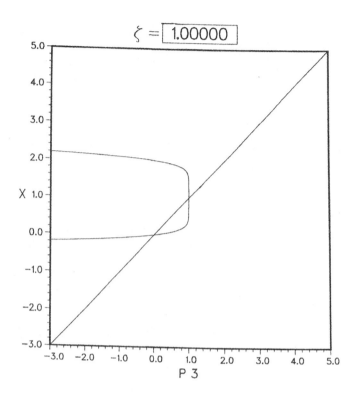

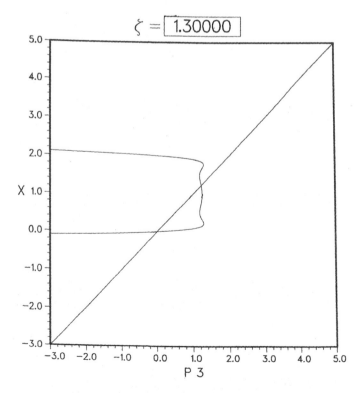

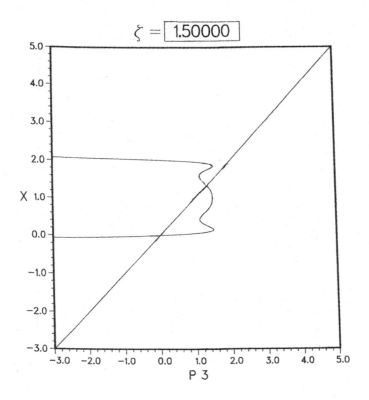

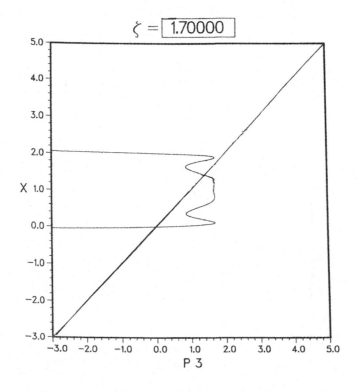

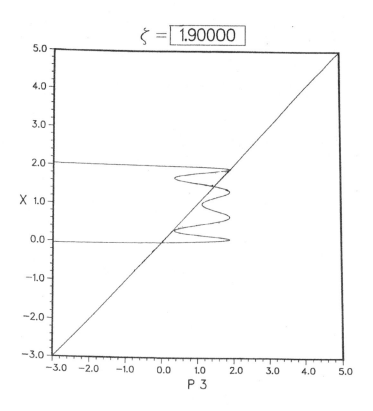

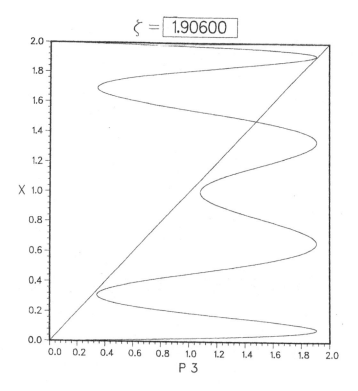

$\zeta = \boxed{1.91200}$

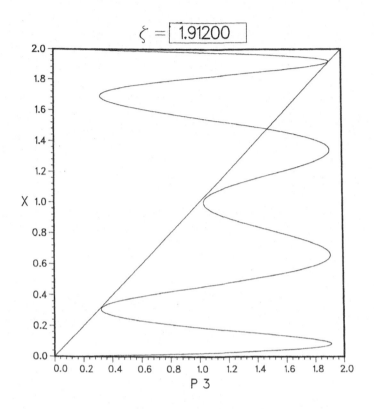

$\zeta = \boxed{1.91400}$

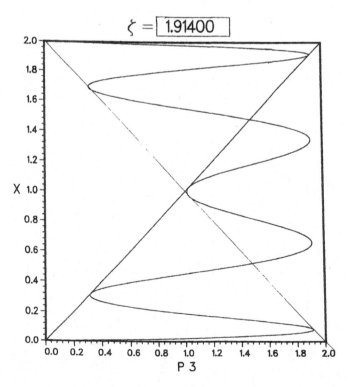

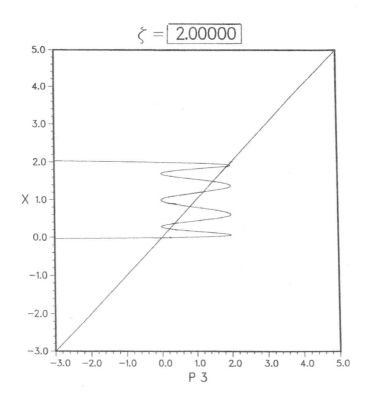

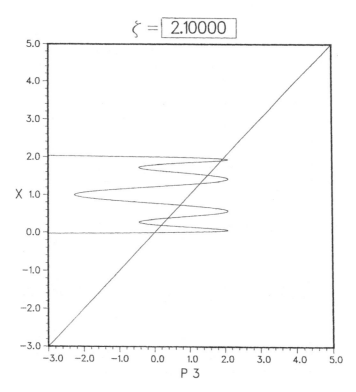

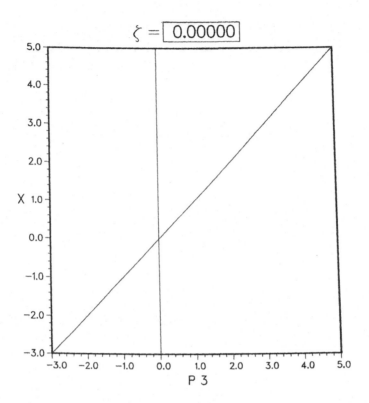

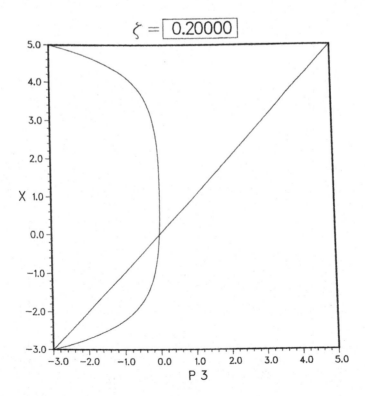

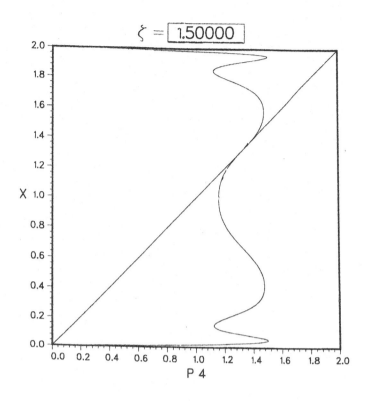

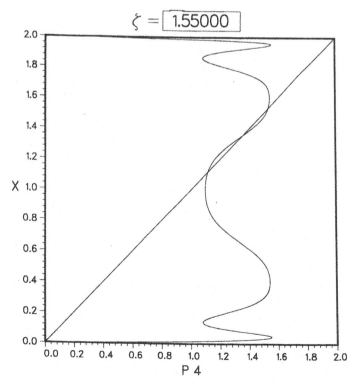

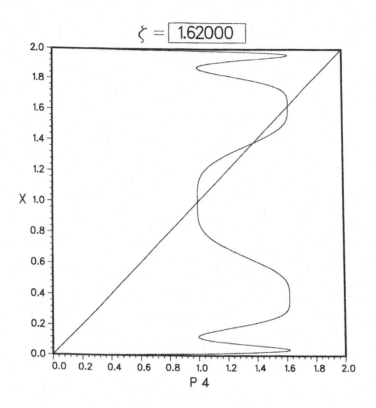

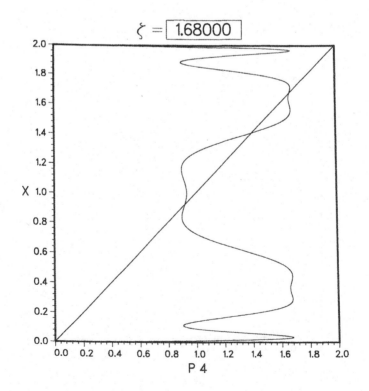

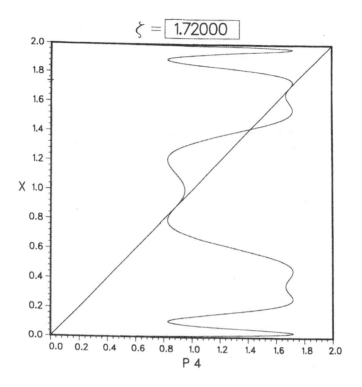

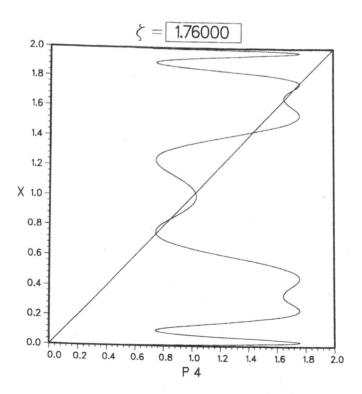

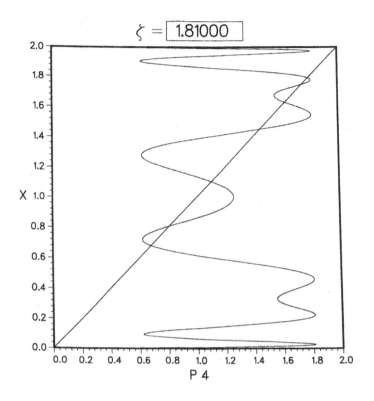

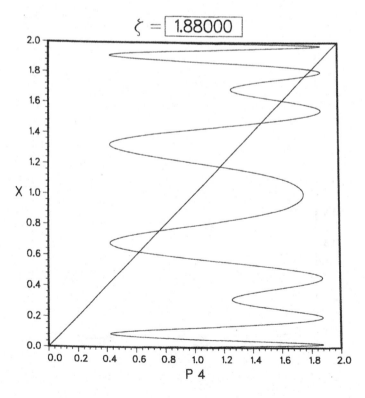

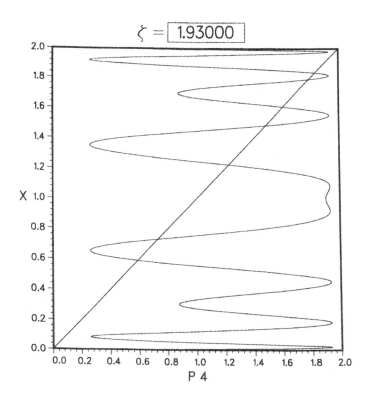

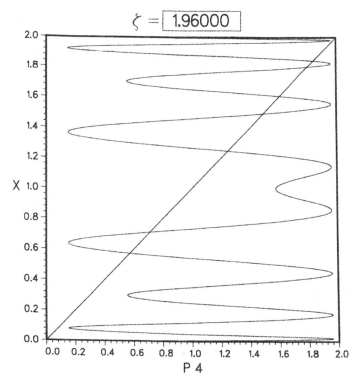

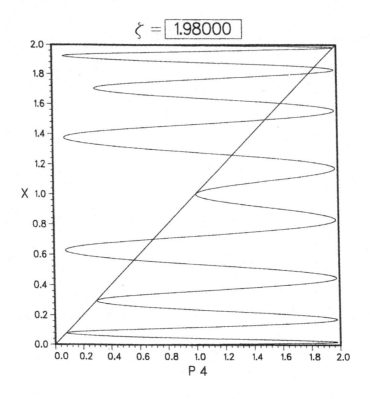

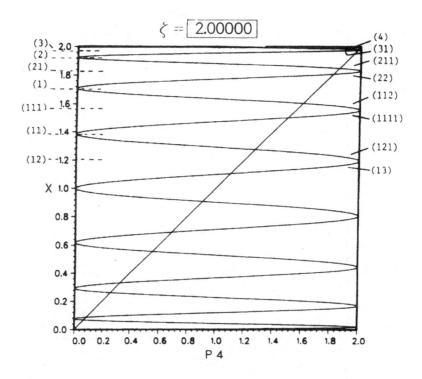

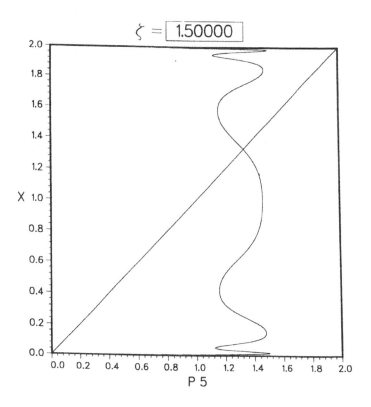

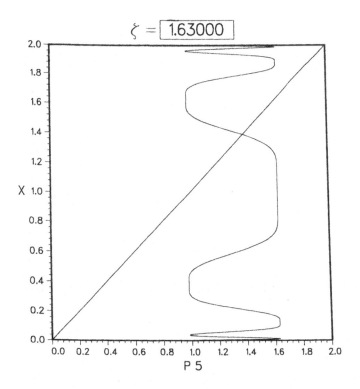

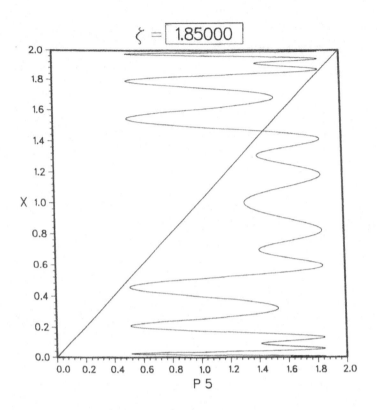

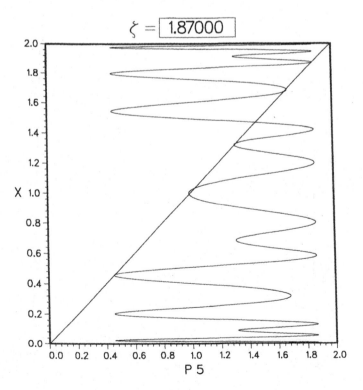

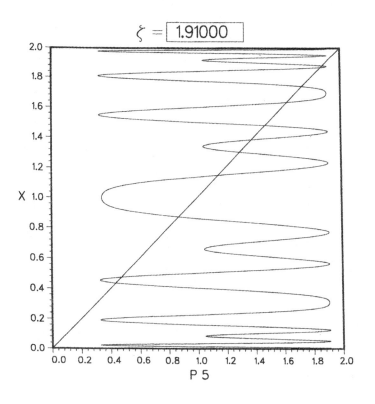

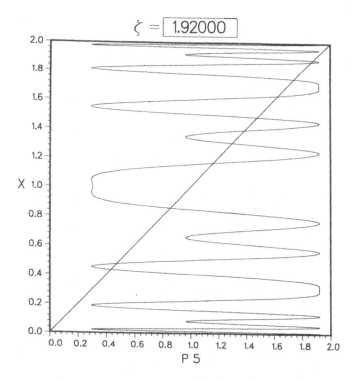

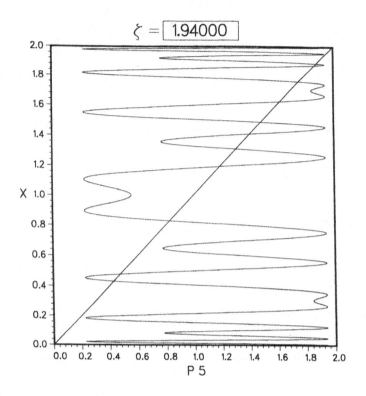

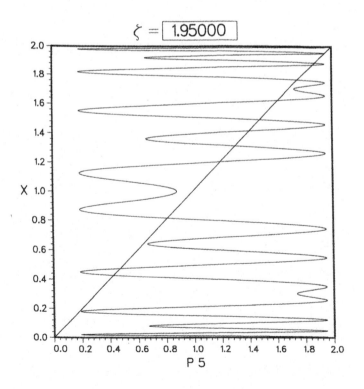

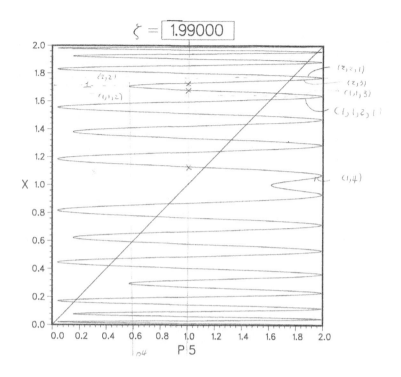

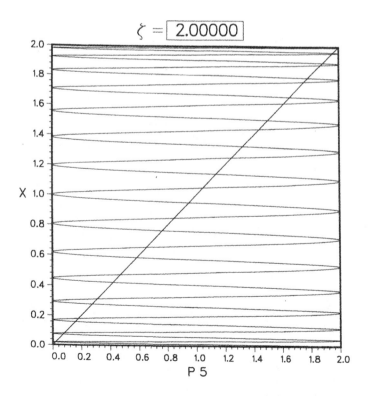

$\zeta = \boxed{2.01000}$

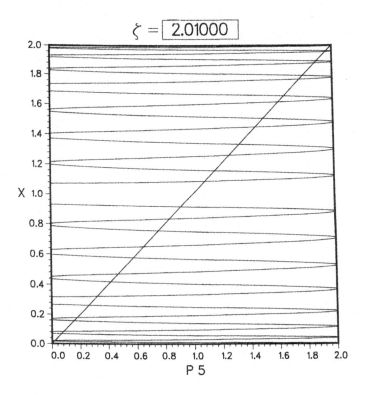

$\zeta = \boxed{1.50000}$

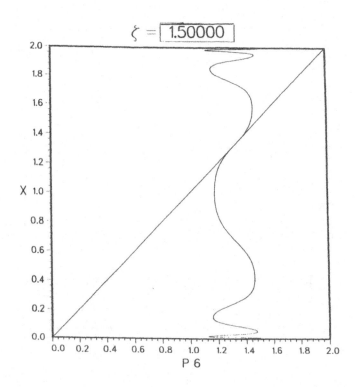

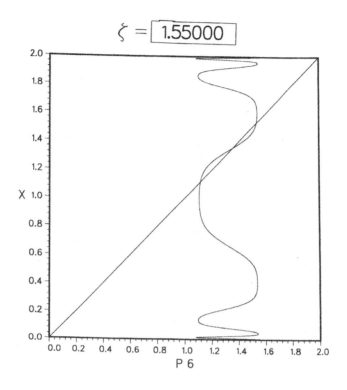

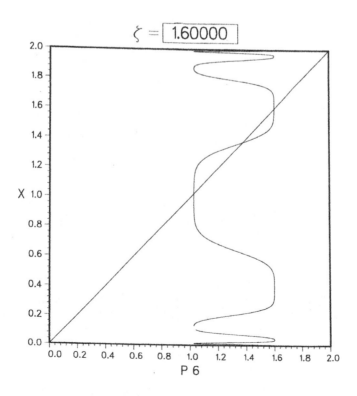

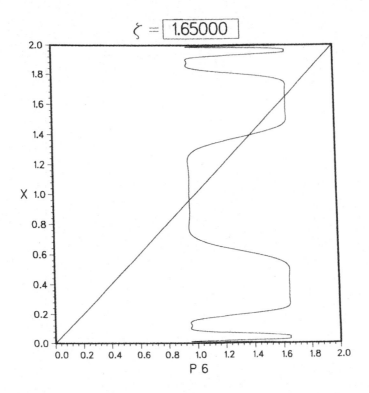

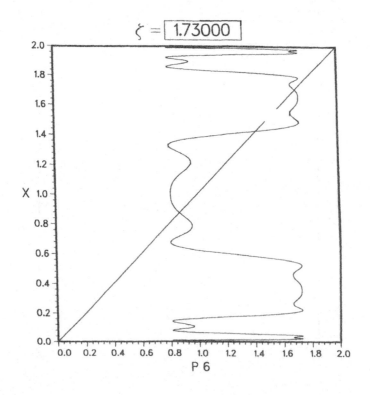

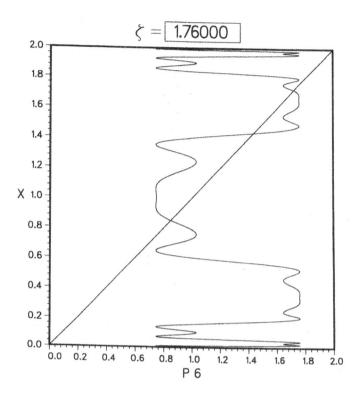

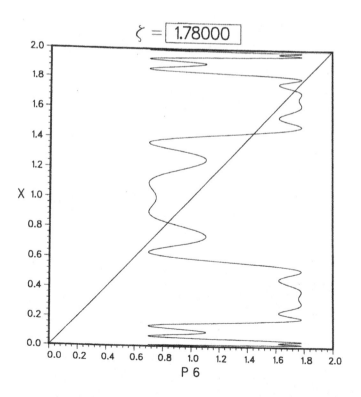

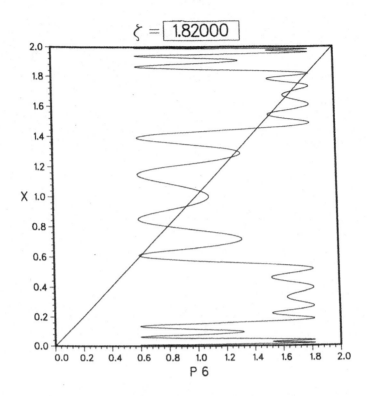

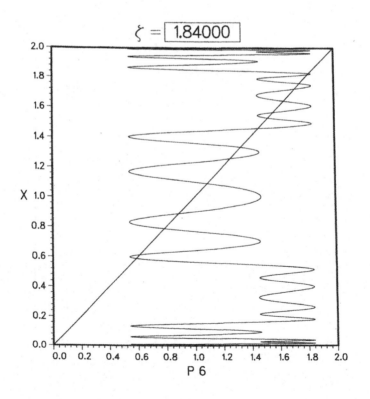

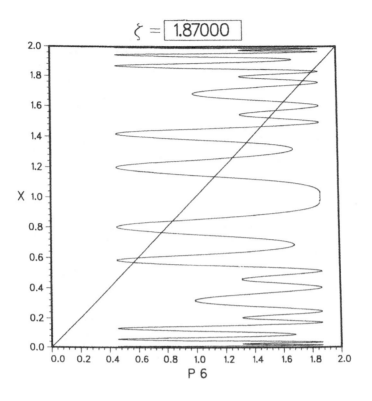

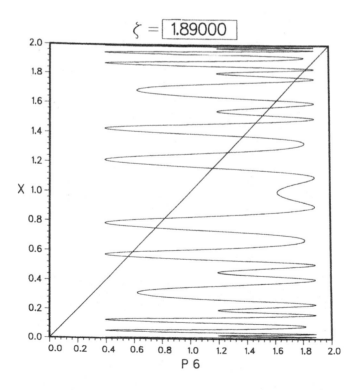

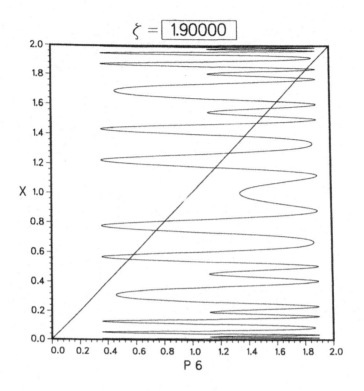

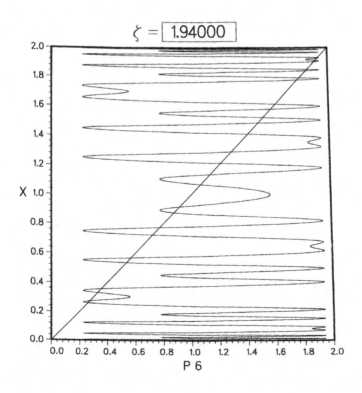

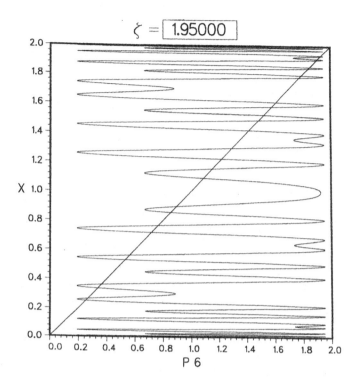

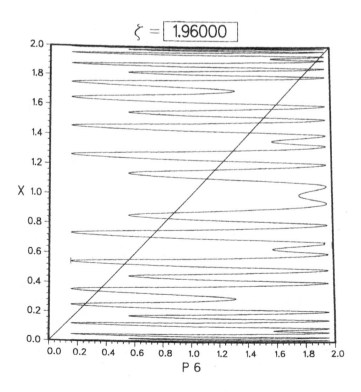

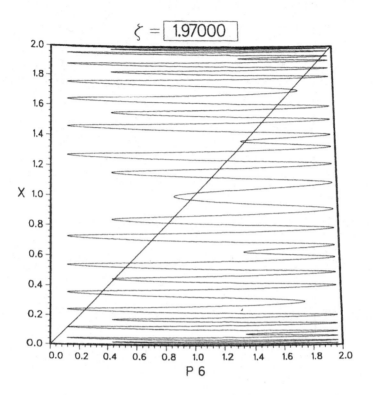

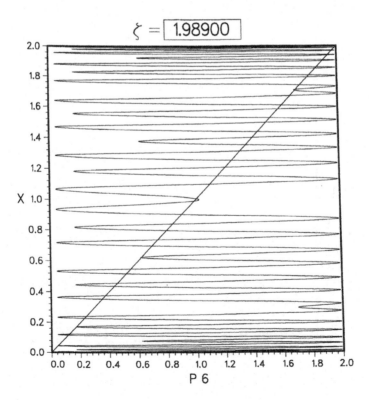

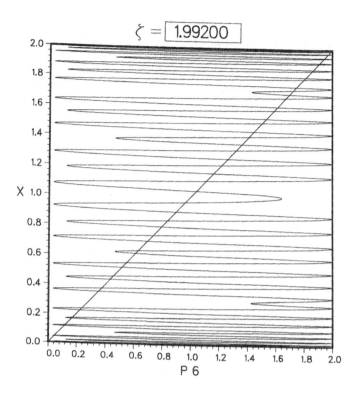

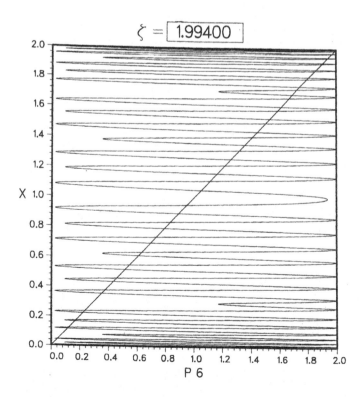

$\zeta = \boxed{1.99500}$

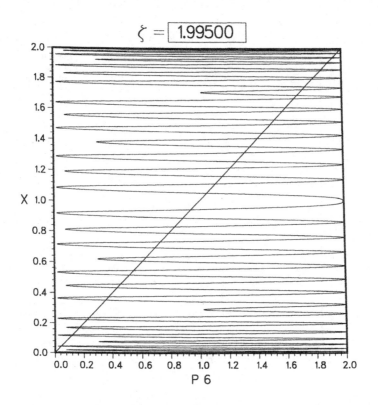

$\zeta = \boxed{1.99600}$

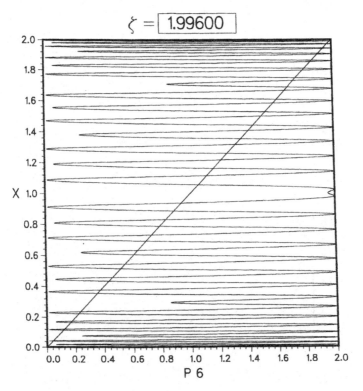

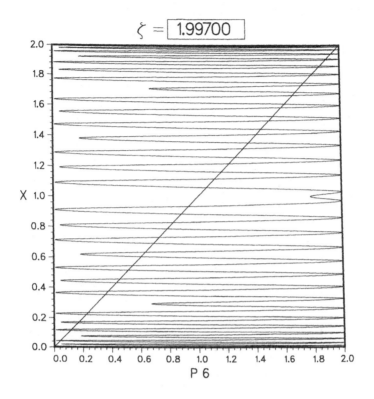

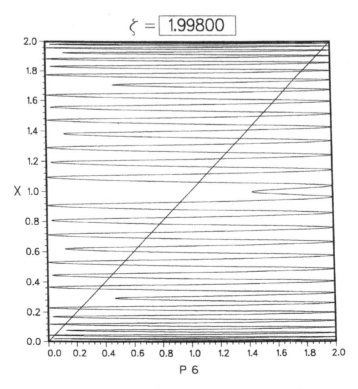

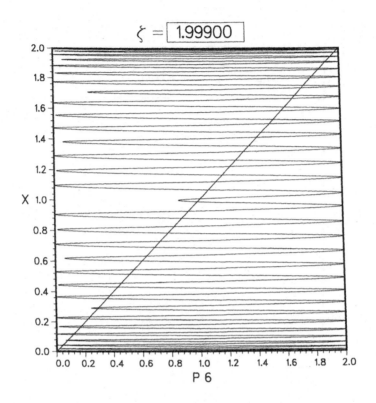

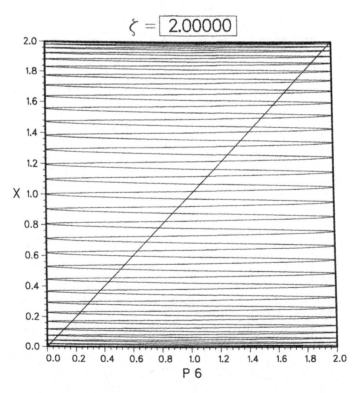

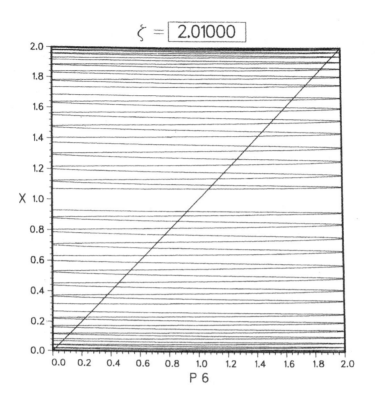

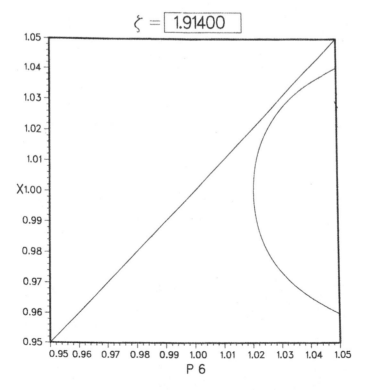

$\zeta = \boxed{1.91500}$

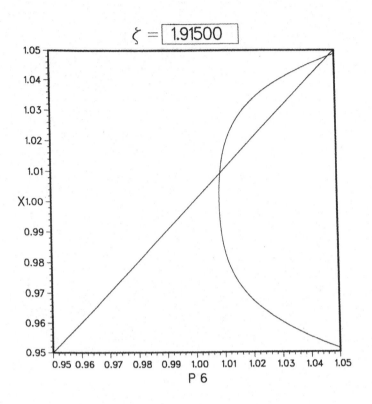

$\zeta = \boxed{1.91600}$

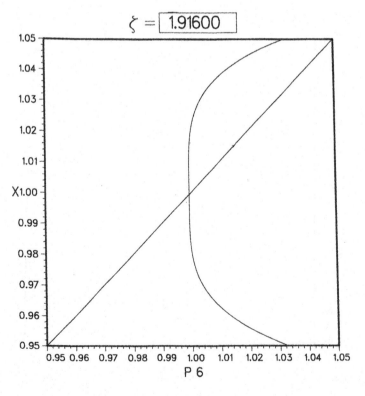

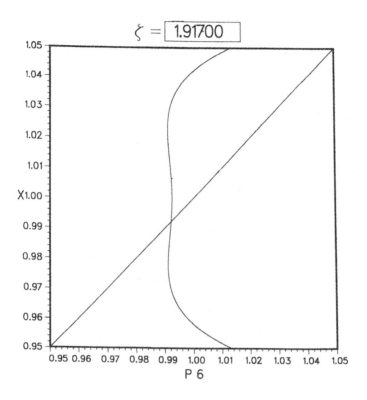

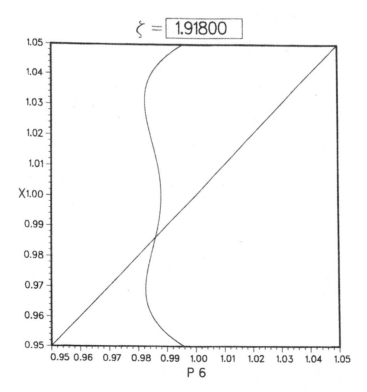

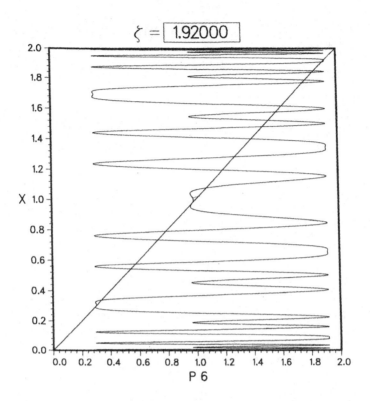

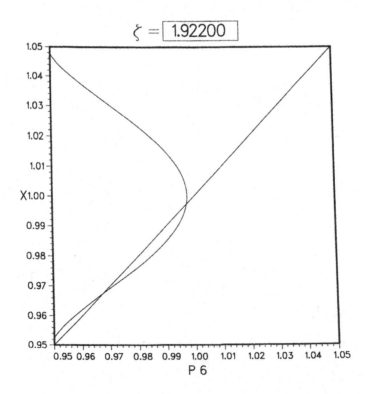

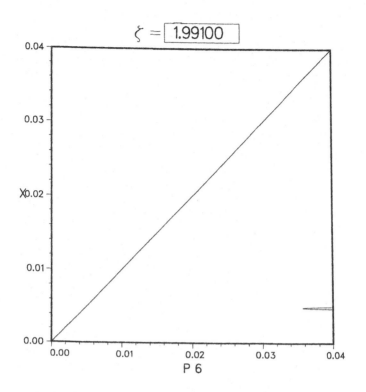

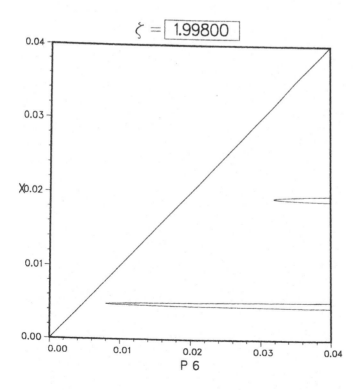

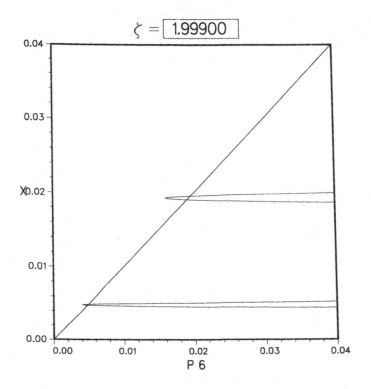

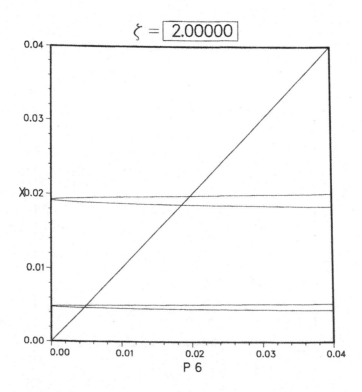

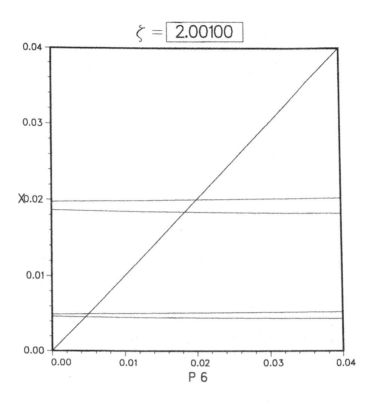

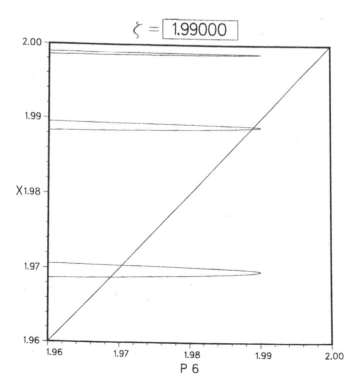

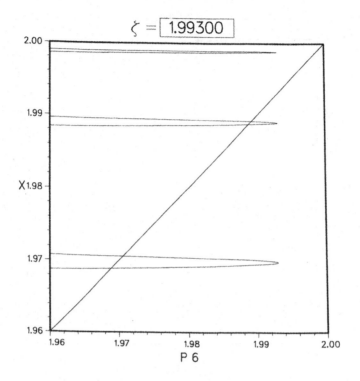

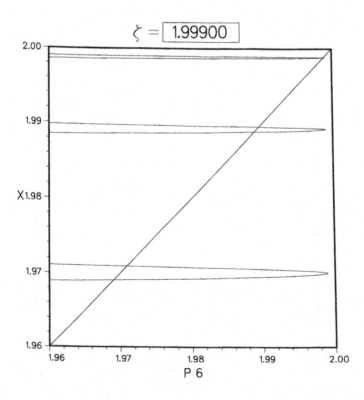

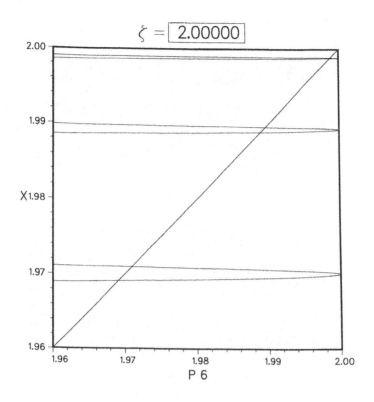

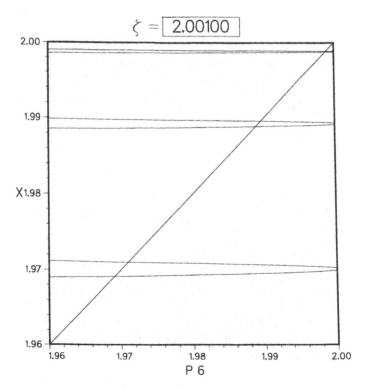

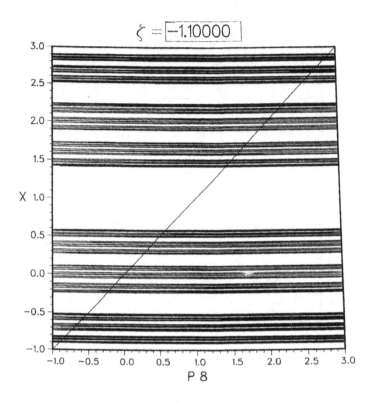

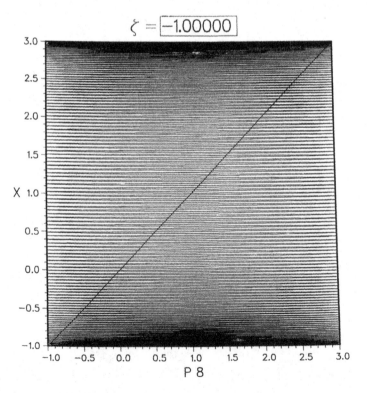

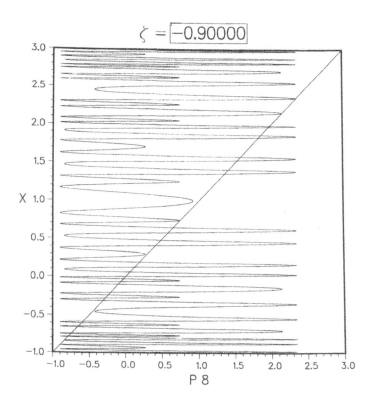

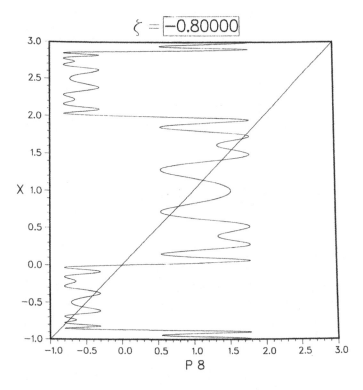

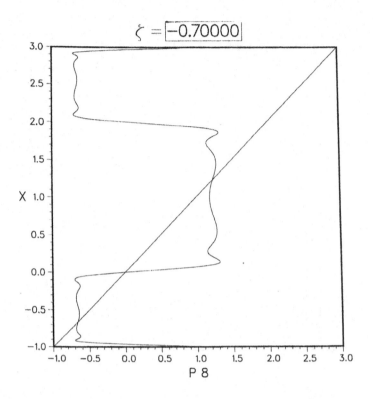

$\zeta = \boxed{-0.70000}$

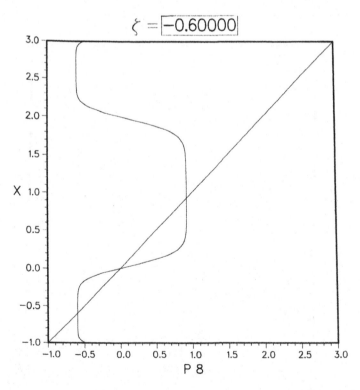

$\zeta = \boxed{-0.60000}$

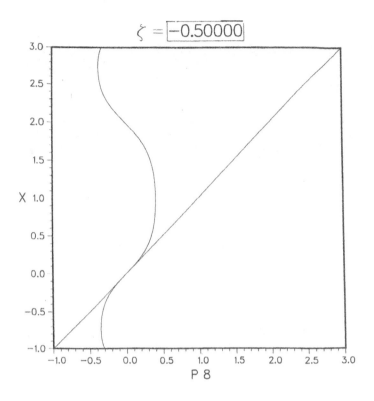

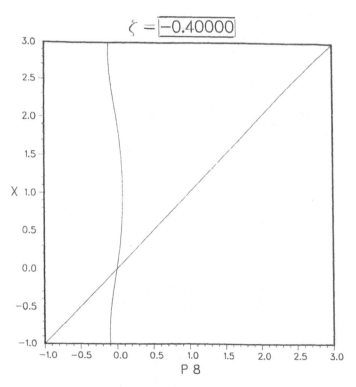

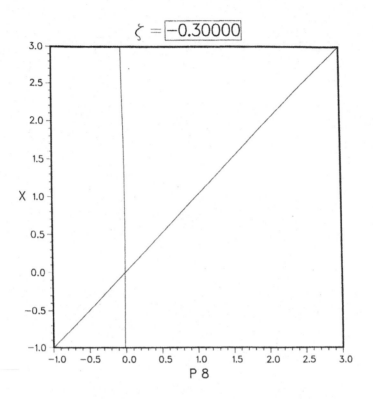

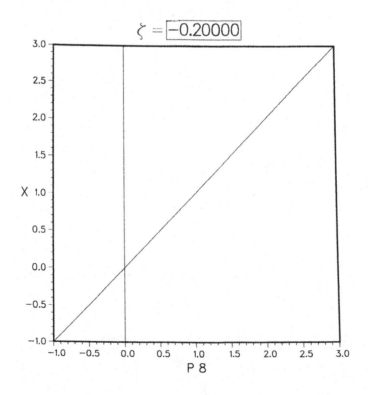

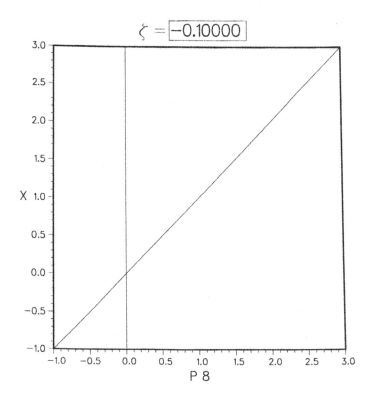

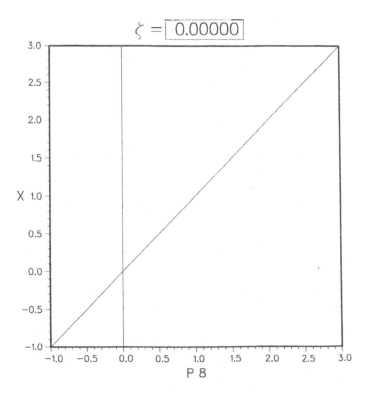

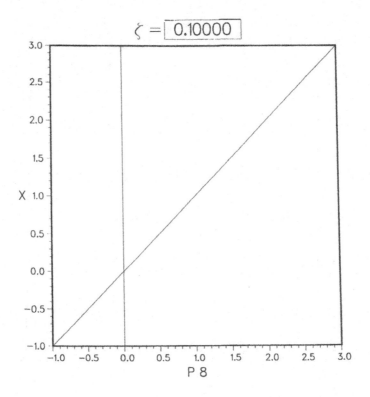

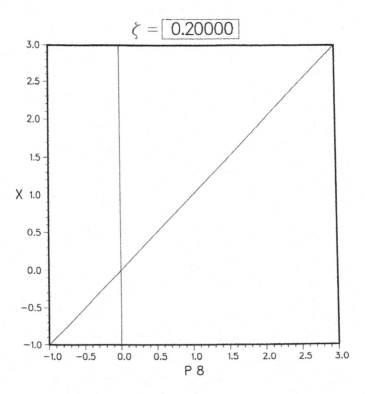

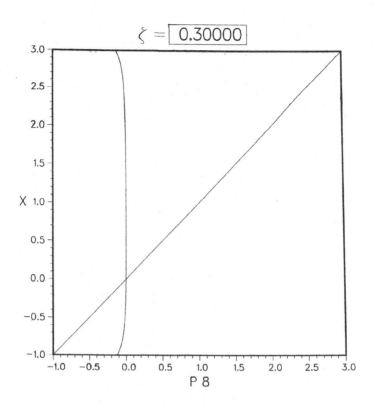

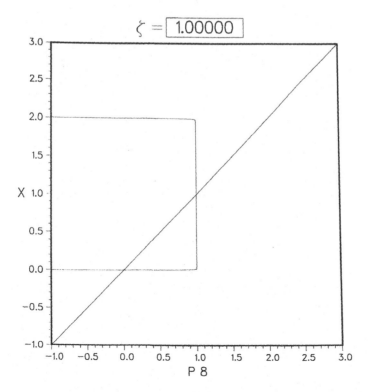

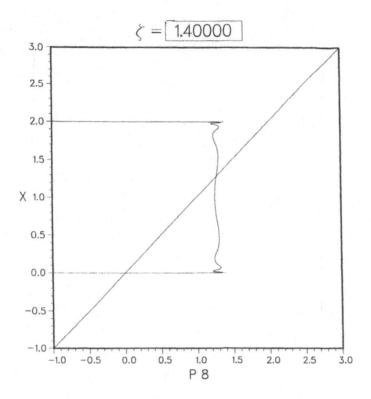

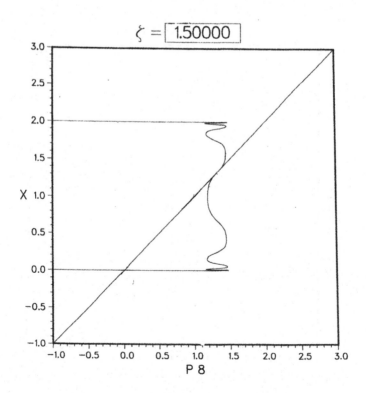

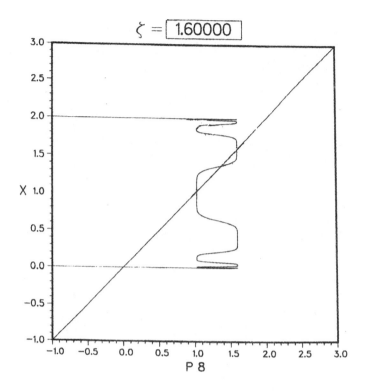

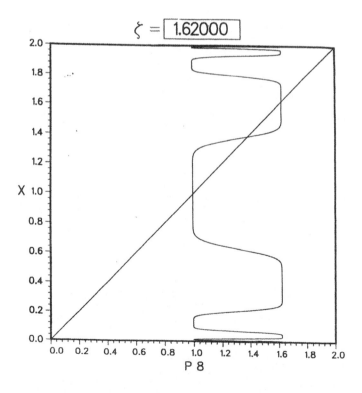

$\zeta = \boxed{1.73000}$

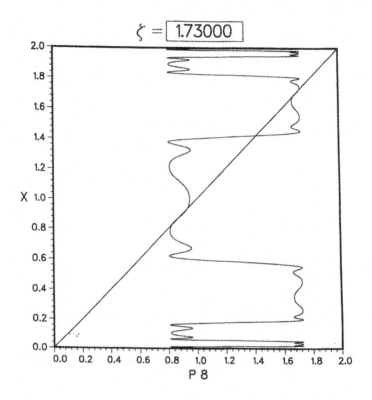

$\zeta = \boxed{1.75800}$

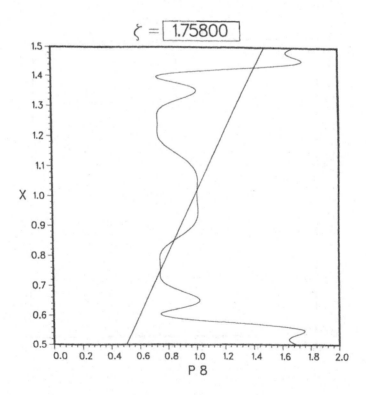

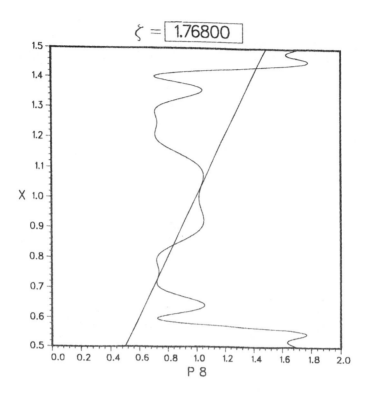

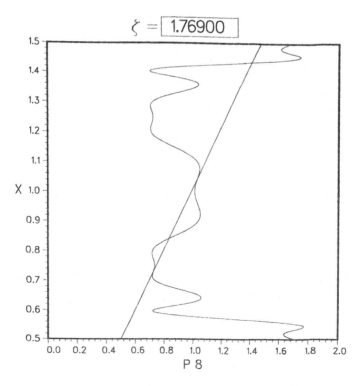

$\zeta = \boxed{1.77100}$

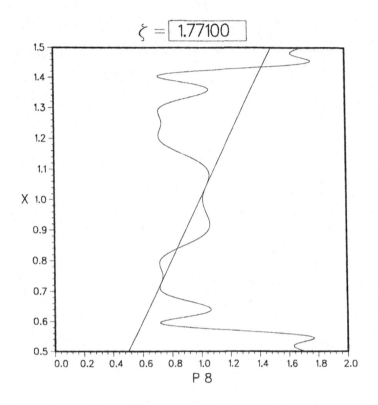

P 8

$\zeta = \boxed{1.78000}$

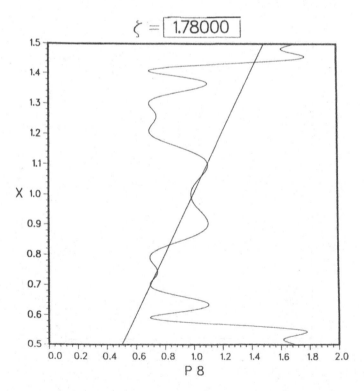

P 8

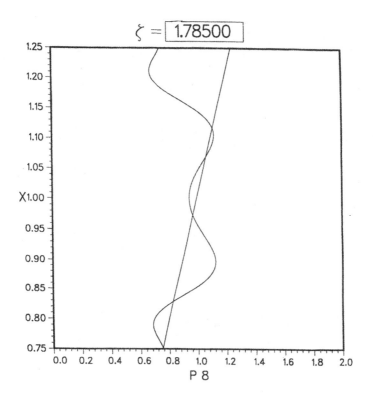

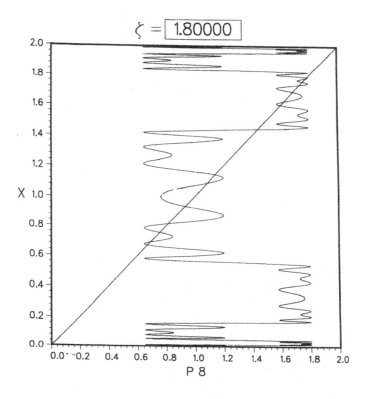

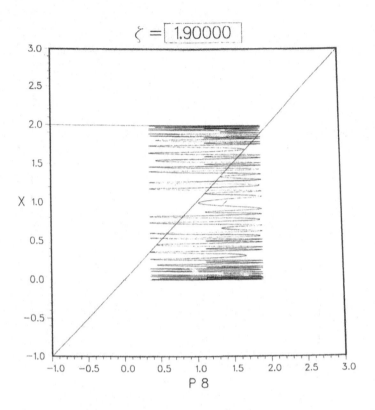

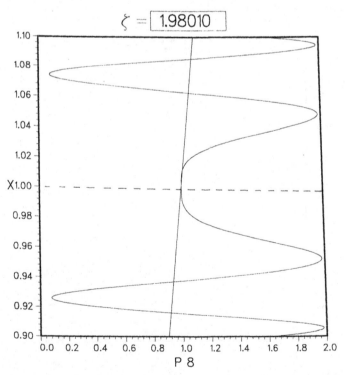

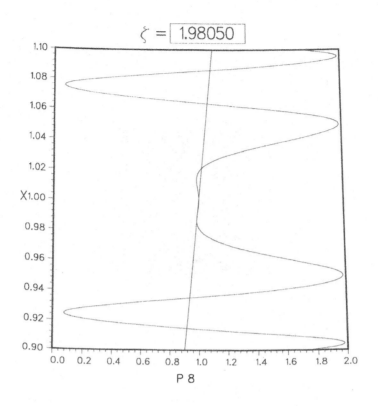

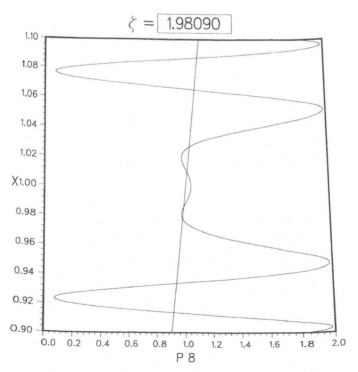

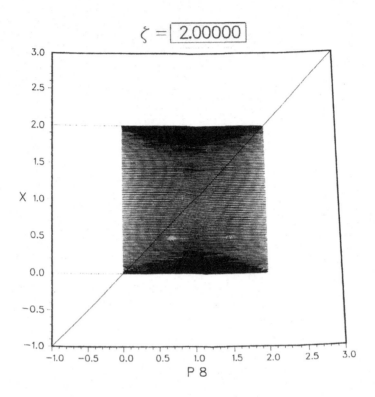

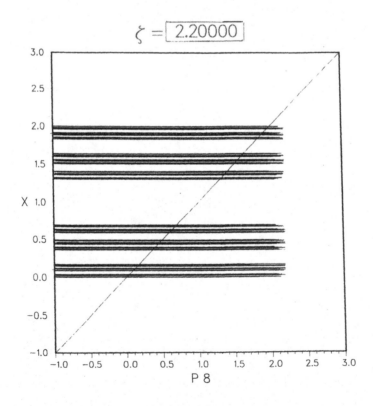

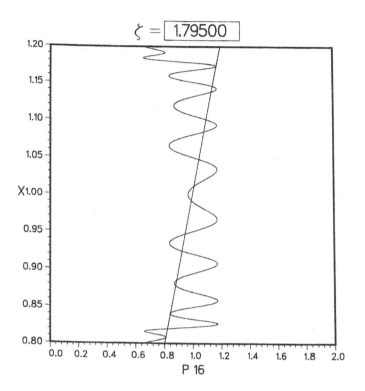

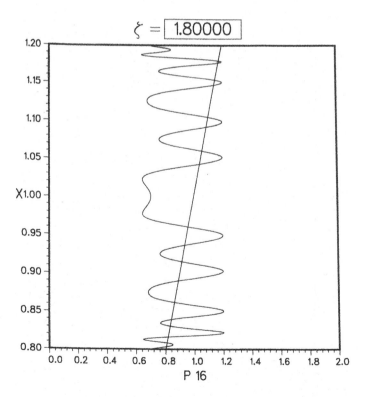

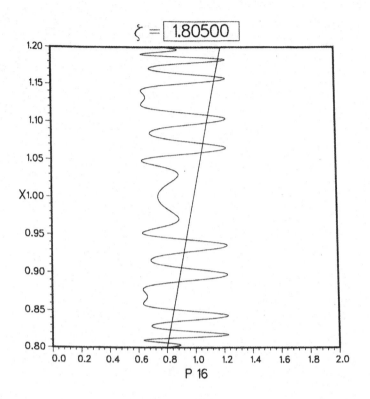

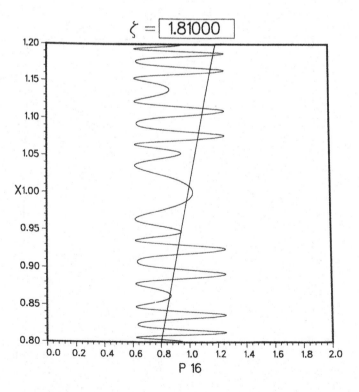

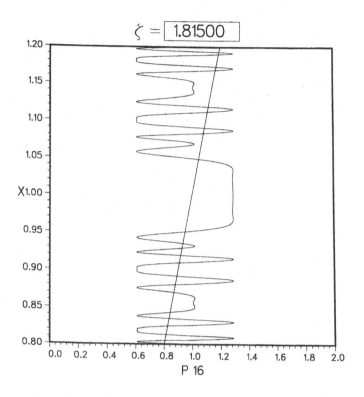

$\zeta = \boxed{1.81500}$

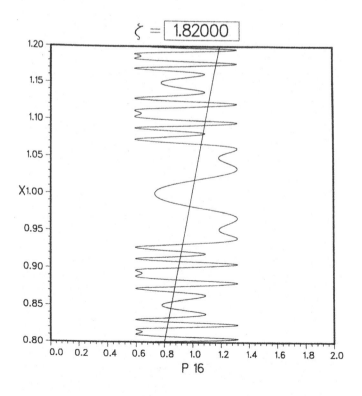

$\zeta = \boxed{1.82000}$

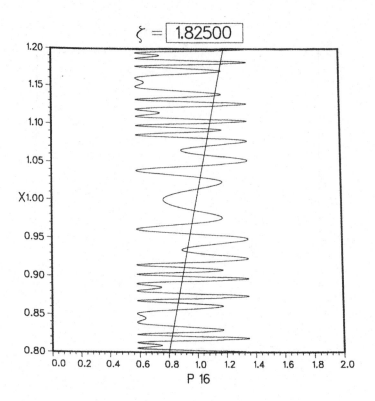

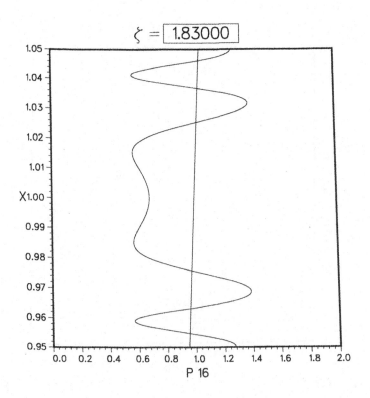

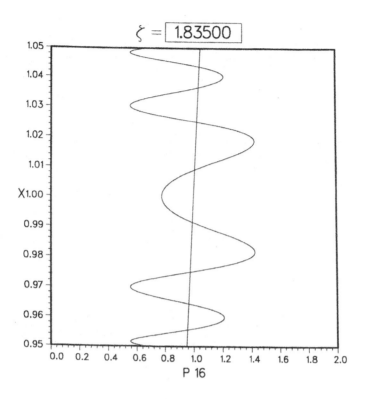

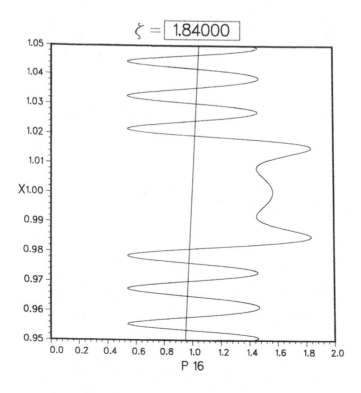

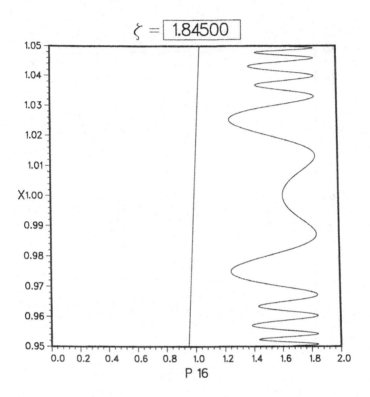

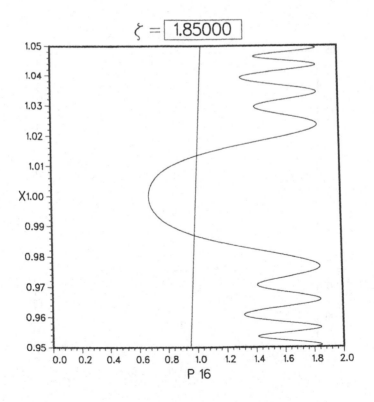

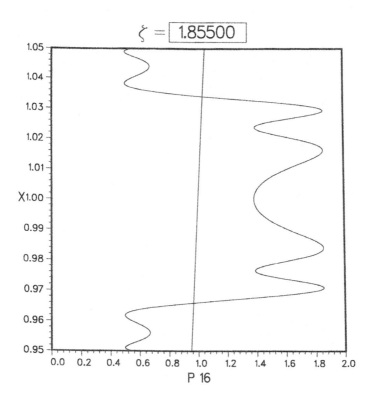

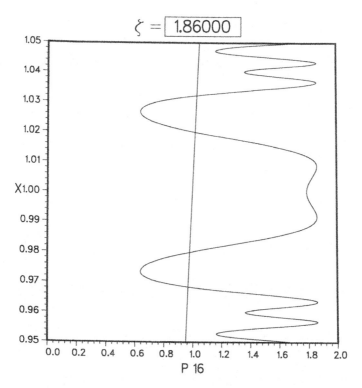

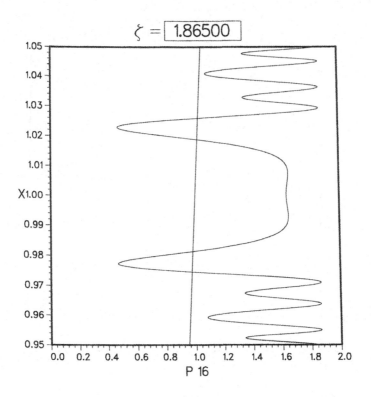

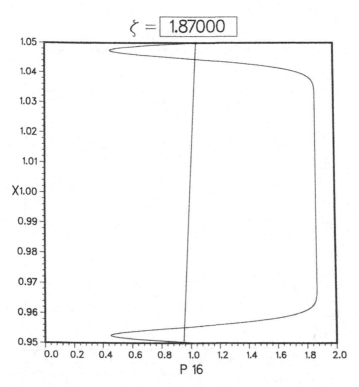

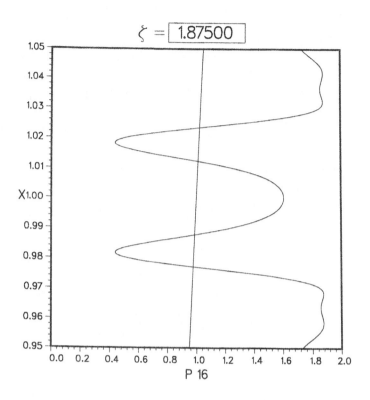

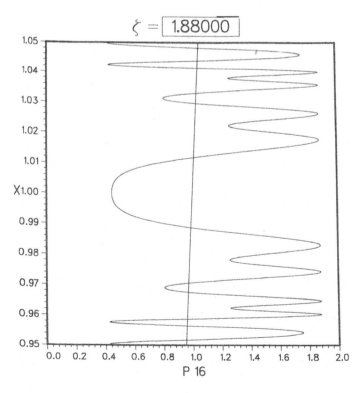

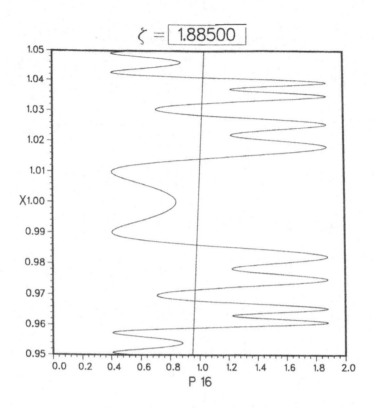

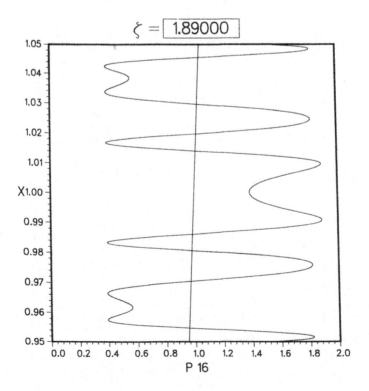

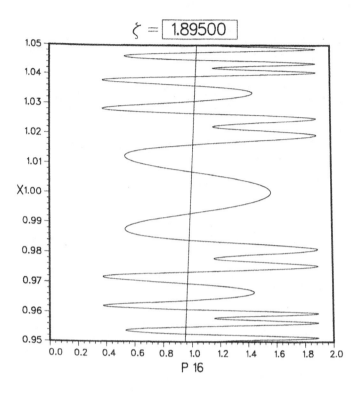

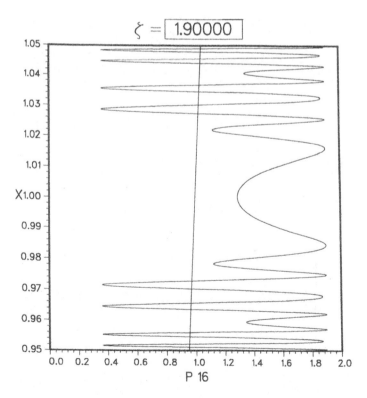

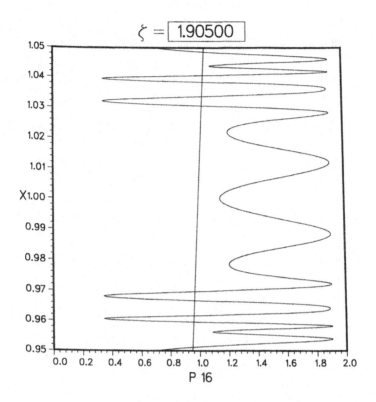

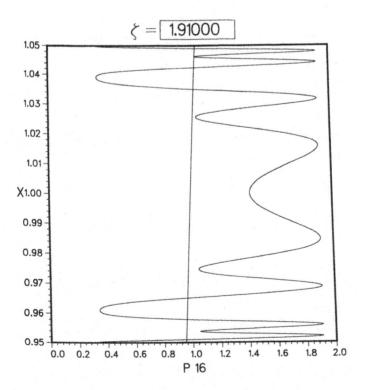

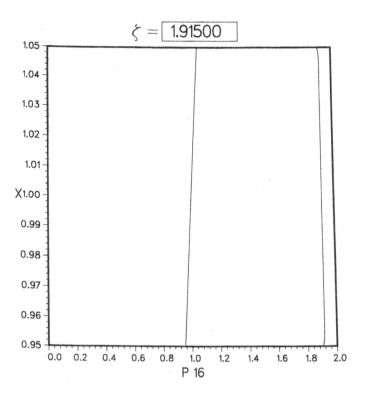

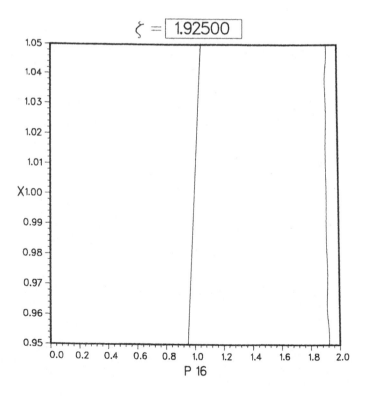

$\zeta = \boxed{1.62000}$

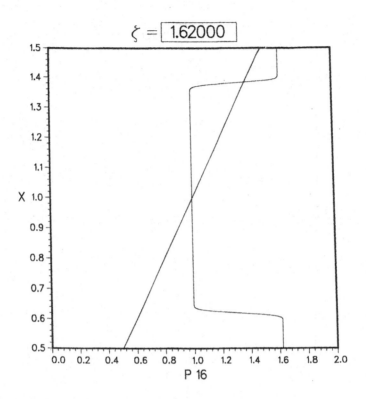

P 16

$\zeta = \boxed{1.72000}$

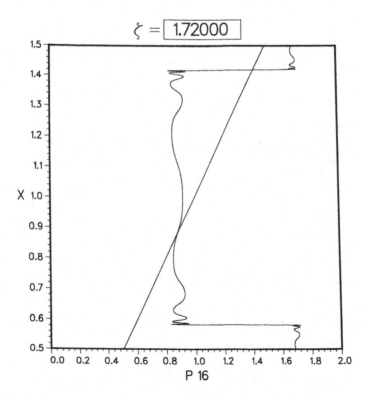

P 16

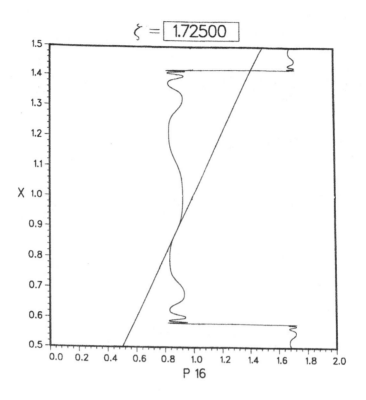

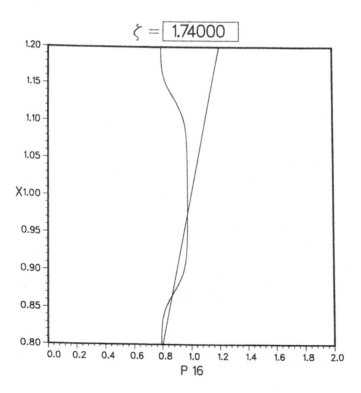

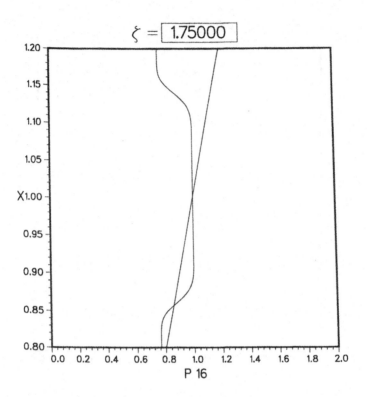

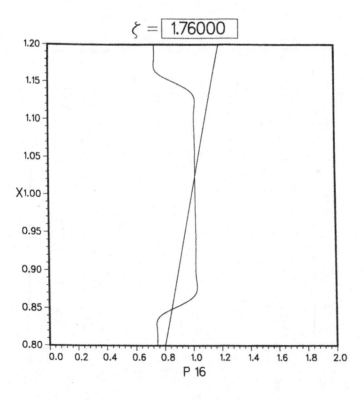

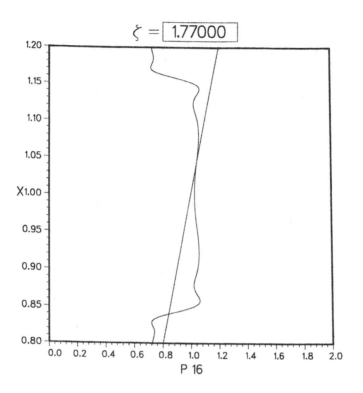

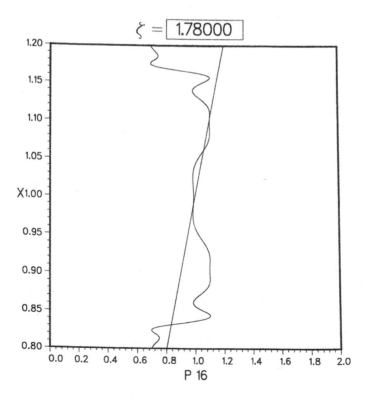

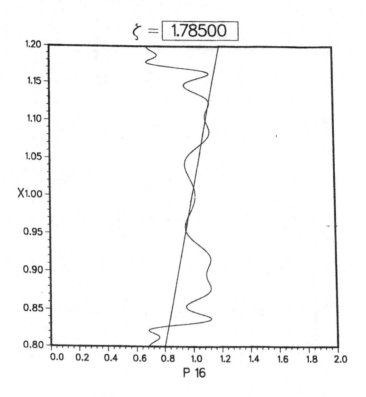

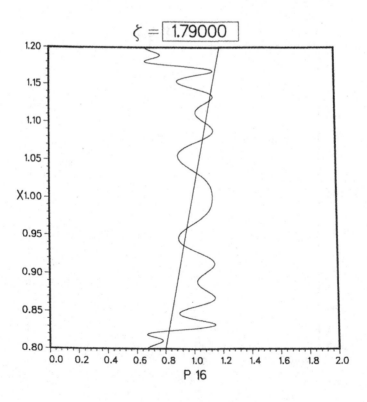

Bibliography

[1] P. R. Stein and S. M. Ulam, Non-linear transformation studies on electronic computers, Rozprawy Matematyczne XXXIX, Warszawa, Panstwowe Wydawnictwo Naukkowe (1964) 1–65. i[1]

[2] N. Metropolis, M. L. Stein, and P. R. Stein, Stable states of non-linear transformations, Numerische Mathematik, **10** (1967) 1–19. i

[3] N. Metropolis, M. L. Stein, and P. R. Stein, On finite limit sets for transformations on the unit interval, J. Combinatorial Theory **15** (1973) 25–44. i

[4] M. J. Feigenbaum, Quantitative universality for a class of nonlinear transformations, J. Stat. Phys. **21** (1978) 25–52. i

[5] M. J. Feigenbaum, The universal metric properties of nonlinear transformations, J. Stat. Phys. **21** (1979) 669–706. i

[6] M. J. Feigenbaum, The onset spectrum of turbulence, Phys. Lett. **74A** (1979) 375–378. i

[7] M. J. Feigenbaum, Metric universality in nonlinear recurrence, Lecture Notes in Physics **93** (1979) 163–166. i

[8] M. J. Feigenbaum, Universal behavior in nonlinear systems, Los Alamos Science (1980) 4–27. i

[9] M. J. Feigenbaum, The transition to aperiodic behavior in turbulent systems, Commun. Math. Phys. **77** (1980) 65–86. i

[10] M. J. Feigenbaum, L. P. Kadanoff, and S. J. Shenker. Quasiperiodicity in dissipative systems: A renormalization group analysis, Physica **5D** (1982 370–386. i

---

[1]The references are presented in a style that relates the references to their usage in this monograph. Some are referenced, some are not. Additional references are included for the convenience of readers of varying backgrounds. Refs. [1]-[28] have their origin with authors having an official relation with the Laboratory. In particular, Refs. [15-19] relate directly to the determination of the properties of the inverse graphs as the primary objects of chaos theory. The applicability of this method to several problems is addressed in Refs. [20] (DNA), [21] (DNA), [22] (DNA), [23] (Galois groups), [24] (Conway numbers), [25] (Conway numbers). The remaining references point out some of the directions in which Chaos Theory has moved in establishing it as a New Branch of Physics. In particular, in their tribute to Lorenz in Ref. [39], Motter and Campbell assess these contributions. It is also appropriate to point out that The Center for Nonlinear Studies was founded at the Laboratory in 1980 with Director Alwyn Scott (1981-1985) and second Director David K. Campbell (1985-1993). The Center continues to this day.

It is hoped that this monograph contributes further to the New Branch of Science features of Chaos Theory by showing that Chaos Theory is a Complex Adaptive System in the sense advocated by the Santa Fe Institute [59].

[11] W. A. Beyer and P. R. Stein, Period doubling for trapezoid function iteration: metric theory, Adv. Appl. Math. **3** (1982) 1–17. **i**

[12] M. J. Feigenbaum, Universal behavior in nonlinear systems, Physica **7D** (1983) 16–39. **i**

[13] W. A. Beyer, M. L. Mauldin, and P.R. Stein, Shift-maximal sequences in function iteration: Existence, uniqueness, and multiplicity, J. Math. Analysis **115** (1986) 305–362. **i**

[14] P. R. Stein, Strange Attractors, and Number Theory, Los Alamos Science (1987) 91–106. **i**

[15] R. L. Bivins, J. D. Louck, N. Metropolis and M. L. Stein, Classification of all cycles of the parabolic map, Physica **D19** (1991) 3–27.

[16] J.D. Louck, Problems in combinatorics on words originating from discrete dynamical systems, Annals of Combinatorics **1** (1997) 99–104.

[17] J. D. Louck, Properties of Metropolis, Stein, and Stein polynomials, LAUR 98-1758 (unpublished) (1998) 1–16.

[18] W. Y. C. Chen, J. D. Louck, J. Wang, Adjacency and parity of words in discrete dynamical systems, J. Comb. Theory, Series **A91** (2000) 476–508.

[19] J. D. Louck and M. L. Stein, Relations between words and maps of the interval, Annals of Combinatorics **5** (2001) 425–449.

[20] G. I. Bell, R. L. Bivins, J. D. Louck, N. Metropolis, and M. L. Stein, Properties of words on four letters from those on two letters with an application to DNA sequences, Adv. Appl. Math. **14** (1993) 348–367.

[21] G. I. Bell and D. C. Torney, Repetitive DNA sequences: Some considerations for simple sequence repeats. Comput. Chem. **17** (1993) 185–190.

[22] W. Y. C. Chen, J. D. Louck, Neckaces, MSS sequences, and DNA sequences, Adv. Appl. Math. **18** (1997) 18–32.

[23] W. A. Beyer and J. D. Louck, Galois groups for polynomials related to quadratic map iterates, Ulam Quarterly **2** (1994) 1–39.

[24] J. D. Louck, Conway numbers and iteration theory, Adv. Appl. Math. **18** (1997) 181–215.

[25] W. A. Beyer and J. D. Louck, Transfinite function induction iteration and surreal numbers, Adv. Appl. Math. **18** (1997) 333–350.

[26] M. T. Menzel, P. R. Stein, and S. M. Ulam, Quadratic Transformations, Part I. Los Alamos National Laboratory Report LA–2305, 1959.

[27] W. A. Beyer and P. R. Stein, Brief History of Functional Iteration at Los Alamos National Laboratory Report LA-9705-H, 1983.

[28] J. D. Louck and N. Metropolis, Symbolic Dynamics of Trapezoidal Maps, Dordrecht, Holland, D. Dreidel Pub. Co. 1986. **i**

[29] S. Smale, Differentible dynamical systems, American Math. Soc. **73** (1967) 747–817.

[30] E. N. Lorenz, The problem of deducing the climate from the governing equations, Tellus **16** (1976) 1–11.

[31] R. M. May, Simple mathematical models with very complicated dynamics, Nature **261** (1976) 459–467.

[32] P. Collet and J.-P. Eckmann, Iterated Transformations of the Interval as Dynamical Systems, Birkhäuser, Boston 1980.

[33] E. N. Lorenz, The predictability of hydrodynamical flow, Transactions of the New York Academy of Sciences Ser. I: **73** (1963) 409–432.

[34] O. Lanford III, A computer-assisted proof of the Feigenbaum conjecture, Bull. Math. Soc. **6** (1982) 427–434.

[35] L. P. Kadanoff, Road to chaos, Phys. Today, December (1983) 46–59.

[36] J. Guckenheimer, Bifurcations of maps of the interval, Invent. Math. **39** (1977) 165–178.

[37] Hao Bai-Lin, Chaos, World Scientific 1984.

[38] D. Lind and B. Marcus, An Introduction to Symbolic Dynamics and Coding, Cambridge University Press 1995.

[39] K. M. Bruchs, Dynamics of One-Dimensional Maps: Symbols, Uniqueness, and Dimension, Ph.D. Dissertation North Texas State University, Denton, Texas, 1988.

[40] Adilson E. Motter and D. K. Campbell, Chaos at 50, Physics Today, May 2013 **66** 27–31.

[41] Abraham Pais, The Science and the Life of Albert Einstein, Oxford University Press 1982.

[42] Paul Davies, God and the New Physics, Simon and Schuster, Inc. New York 1984.

[43] Alvin Powell, A Theory about Everything, Harvard University Gazette 2/10/2000 (on the work of Juan Maldacena).

[44] Ron Cowen, Is the Universe a Hologram?, Nature, posted 12/12/2013 (on the work of Juan Maldacena).

[45] Carl Zimmer, The Secrets of the Brain, National Geographic **225** 2004 (on the work of Van Wedeen and Jeff Lichtman).

[46] James Owen Weatherall, The Physics of Wall Street, Houghton Mifflin Harcourt Pub. Co, New York 2013 (contains information on the contributions of Benoit Mandelbrot and James Farmer, the latter now at the Santa Fe Institute).

[47] E. P. Wigner, The unreasonable effectiveness of mathematic in the natural sciences, Comm. Pure and Appl. Sciences **13** (1960) 1–14.

[48] Eugene P. Wigner, The problem of measurement, Am. J. Phys. **31** (1963) 6–15.

[49] Eugene P. Wigner, On hidden variables and quantum mechanical probabilities, Am. J. Phys. **38** (1970) 1005–1009.

[50] Lee Smolen, Time Reborn: From the Crisis in Physics to the Future of the Universe, Houghton Mifflin Harcourt 2013.

[51] Lee Smolen, Time, laws, and the future of cosmology, Physics Today **67**(3) 38–43.

[52] A. E. Motter and D. K. Campbell, Physics Today **67**(3) 9–10 (Letter by David Ruelle with response).

[53] (a) James D. Louck, Unitary Symmetry and Combinatorics, World Scientific, Singapore 2008; (b) Applications of Unitary Symmetry and Combinatorics 2011.

[54] Richard P. Stanley, Enumerative Combinatorics Vol 1, Cambridge University Press 1997; Vol 2 1999.

[55] L. C. Biedenharn and J. D. Louck, Angular Momentum in Quantum Physics, Cambridge University Press 1985 (digital 2009).

[56] V. Dolotin and A. Morozov (a) Introduction to Non-Linear Algebra, World Scientific 2007 (b) Universal Mandelbrot Set, World Scientific 2006.

[57] Stephen Hawking and Roger Penrose, The Nature of Space and Time, Princeton Science Press 1996 (and references therein).

[58] Santa Fe Institute, 1399 Hyde Park Road, NM, 87501, USA **87**.

Epilogue

This monograph is not an historical account of chaos theory; what has been added to earlier chaos theory is that

Chaos Theory is a Complex Adaptive System under the operation of function composition as evidenced by its unique algorithmic-computer-generated $\mathbb{C}_n^*$ construction.

This raises the possibility that the following collection of objects is each a Complex Adaptive System. This is because each of these sets of objects has long been known (see Bibliography) to have some properties that fall under the purview of chaos theory. Here this purview is extended to **all** properties of each of these sets of objects:

1. DNA
2. WEATHER
3. CONWAY NUMBERS
4. GALOIS GROUPS

All items mentioned are on chaos theory originating from Los Alamos National Laboratory. But, of course, this list can be extended to a very long list of topics by consulting the recent review article by Motter and Campbell [40] and the journal *Chaos Theory* that establishes chaos theory as a new science—a new way of viewing physical systems. This, of course, is also the view developed in this monograph and carrying it a step further, perhaps, by suggesting that many such systems showing behavior described by chaos theory might also be considered as complex adaptive systems. The application to General Relativity is of particular interest, especially in view of the existence of a unique, completely deterministic chaos theory. Of course, this property can now be applied also to each of the sets of objects above.

Abraham Pais [41] puts forth a powerful portrait of Einstein whose primary goal was to show that general relativity would eventually be shown to imply quantum theory. In pursuit of that goal he considered all other reasonable approaches and criticisms through personal audiences and many correspondences. His enormous commitment to good science was unmatched; it cannot be captured here in a few lines. In addition, he was very much aware of events outside of science that were changed by his new view of the Universe, including religion, and gave his considered advice on what it meant. The presentation of General Relativity as a Complex Adaptive System given in this monograph is made in the same good-science spirit.

1. The book by Weatherall [46] is a delightful journey through much of modern physics, including quantum theory and general relativity. But the book presents a radical concept of how the world operates through a new agency called the Theory of Modeling, which has foundations in Economics. It is a sort of *A Theory of Everything*, but appears to have the shortcoming of giving no details on implementation. It seems, however, to be consistent with the viewpoint that such systems are complex adaptive systems.

2. Many physicists take the viewpoint that physics is based on laws of nature that are universal, laws that do not fall under the same rules as economics and Wall Street protocols and that the extension by Weatherall to physical systems is flawed. Wigner [47] gives an in-depth discussion of the concept "law of nature" from which it is clear that it works, until it does not. This concept is too fragile to be taken as a basic rule for the foundations of physical phenomenon, unless shown otherwise. (See Smolen [50-51] for discussion of this.) In particular, it is surely consistent for Weatherall to include physics in his Theory of Modeling.[1]

3. Closely related to the discussions in Items (1) and (2) is the notion that one of the purposes of mathematics is to explain the behavior of physical systems. This is, no doubt, the case. But physical systems do what they do with no input from mathematics, which has a primary use of bring some organization to our experiences. Mathematics has many purposes, not all of which relate to physics. In particular, see Wigner [48-49] for an interpretation of the meaning of measurement in quantum mechanics that bears as well on the present topic. In particular, as Wigner shows, the concept of "natural law" is too fragile to be taken as a basic concept in physical laws.

4. A criticism of the applications of the methods advanced in this monograph is that they cannot be extended in an obvious way to three spatial dimensions. It may be possible to gain some understanding of the reason for this by assuming that mathematics is, by its very nature, not capable of providing such insights: The best that can be done is to create models that help organize in some reasonable way what is going on in complex physical systems.[2]

5. Carl Zimmer [45] attributes to Romón y Cajal the insight: "that each neuron is a distinct cell, separate from every other one." This opens up the possibility that the collection of all neurons in every human being is a complex adaptive system.

---

[1] My good friend Don Hansen introduced me to the book by Weatherall.
[2] Discussions with Doron Zeilburger on computer proofs were very useful here.

Index

Printed in the United States
By Bookmasters